GENERALIZATIONS OF PY THEOREM TO POLYGONS

Ram Bilas Misra

CWP
Central West Publishing

GENERALIZATIONS OF PYTHAGORAS THEOREM TO POLYGONS

by

Prof. Dr. Ram Bilas Misra

Ex Vice Chancellor, Avadh University, Faizabad, U.P. (India);

***Professor* of Mathematics, Research & Strategic Studies Centre,**

Lebanese French University, Erbil, Kurdistan (Iraq).

Former: *Dean*, Faculty of Science, A.P. Singh University, Rewa, M.P. (**India**);
Prof., Dept. of Maths., Higher College of Edn., Aden Univ., Aden (**Yemen**);
Professor & Head, Dept. of Maths. & Stats., A.P.S. University, Rewa, M.P. (**India**);
Prof., Dept. of Maths., College of Science, Salahaddin University, Erbil (**Iraq**);
UGC Visiting Prof., Mahatma Gandhi Kashi *Vidyapith*,Varanasi, U.P. (**India**);
Professor, Dept. of Maths, Ahmadu Bello Univ., Zaria (**Nigeria**) – designate;
Prof. & Head, Dept. of Maths. & Comp. Sci., Univ. of Asmara, Asmara (**Eritrea**);
Director, Unique Inst. of Business & Technol., Modi Nagar, Ghaziabad, U.P. (**India**);
Prof. & Head, Dept. of Maths., Phys. & Stats., Univ. of Guyana, Georgetown (**Guyana**);
Prof. & Head, Dept. of Maths., Eritrea Inst. of Technology, Mai Nefhi (**Eritrea**);
Prof.& Head, Dept. of Maths., School of Engg., Amity Univ., Lucknow, U.P. (**India**);
Prof. & Head, Dept. of Maths. & Comp. Sci., PNG Univ. of Technology, Lae (**PNG**);
Prof. of Maths., Teerthankar Mahaveer University, Moradabad, U.P. (**India**);
Prof., Dept. of Maths, Oduduwa Univ., Ipetumodu, Osun State (**Nigeria**) – designate;
Prof., Dept. of Maths, Adama Science & Technology Univ., Adama (**Ethiopia**);
Prof. & Head, Dept. of Maths. & C.S., Bougainville Inst. of Bus. & Tech., Buka (**PNG**);
Prof. & Head, Dept. of Maths., J.J.T. University, Jhunjhunu, Rajasthan (**India**);
Dean, Faculty of Science, J.J.T. University, Jhunjhunu, Rajasthan (**India**);
Professor, Dept. of Maths, Wollo University, Dessie, Wollo (**Ethiopia**);
Professor, Dept. of Appld. Maths., State Univ. of New York, Incheon (**S. Korea**);
Prof., Dept. of Maths. & Computing Sci., Divine Word Univ., Madang (**PNG**);
Director, Maths., School of Sci. & Engg., Univ. of Kurdistan Hewler, Erbil (**Iraq**);
DAAD Fellow, University of Bonn, Bonn (**Germany**);
Visiting Professor, University of Turin, Turin (**Italy**);
Visiting Professor, University of Trieste, Trieste (**Italy**);
Visiting Professor, University of Padua, Padua (**Italy**);
Visiting Professor, International Centre for Theoretical Physics, Trieste (**Italy**);
Visiting Professor, University of Wroclaw, Wroclaw (**Poland**);
Visiting Professor, University of Sopron, Sopron (**Hungary**);
Reader, Dept. of Maths. & Stats., South Gujarat University, Surat, Gujarat (**India**);
Reader, Dept. of Maths. & Stats., University of Allahabad, Allahabad, U.P. (**India**);
Asst. Prof., Dept. of Maths., College of Sci., Mosul Univ., Mosul (**Iraq**) – designate;
Senior most *NCC Officer* (Naval Wing), Univ. of Allahabad, Allahabad, U.P. (**India**);
Lecturer, Dept. of Maths., KKV Degree College, Lucknow, U.P. (**India**).

2020

This edition has been published by Central West Publishing, Australia

For more information about the books published by Central West Publishing, please visit https://centralwestpublishing.com

A catalogue record for this book is available from the National Library of Australia

ISBN (print): 978-1-925823-81-3

DEDICATED TO

My Mathematics Teachers & Relatives

Shri Abdul Lateef
(Primary School, Semrai & Junior High School, Gola Gokarannath);

Shri Abdul Azeez
(Junior High School, Gola Gokarannath);

Shri Ram Nath Singh
(Public Intermediate College, Gola Gokarannath);

Shri Vijay Bahadur Shrivastava
(Public Intermediate College, Gola Gokarannath);

Shri R.S. Pathak
(Kanya Kubja Degree College, Lucknow);

Prof. Dr. Ram Ballabh
(Head, Department of Mathematics & Astronomy, Lucknow University, Lucknow);

Prof. Dr. Ratna Prakash Agrawal
(Department of Maths. & Astronomy, Lucknow University, Lucknow);

Prof. Dr. J.P. Jaiswal
(Department of Maths. & Astronomy, Lucknow University, Lucknow);

Prof. Dr. Amar Nath Mehra
(Department of Maths. & Astronomy, Lucknow University, Lucknow);

Prof. Dr. Hari Mohan Shrivastava
(Department of Maths. & Astronomy, Lucknow University, Lucknow);

Shri Mahadev Prasad Misra, Singahi
(Principal, Govt. Intermediate College, Bijnor);

Shri Basant Lal Misra *– my uncle*
(The First elected Village Pradhan, Semrai);

Shri Hriday Narain Mishra *– the most affectionate relative*
(Panchayat Secretary, Kumbhi Block, Gola Gokarannath);

Shri Ram Kumar Awasthi "Laloose"
(A cousin from the maternal family of grandmother).

CONTENTS

PREFACE

The present book is the outcome of the author's own results discovered recently. It deals with the generalizations of Pythagoras theorem to polygons. The celebrated result of the Pythagoras theorem representing the sum of squares of two (positive) integers as the square of another integer has been extended to quadrilaterals composed of two right triangles so that the sum of squares of its first three sides equals the square of the remaining side. In the language of algebra, integral solutions of a quadratic equation $a^2 + b^2 + c^2 = d^2$ are explored. The first 18 Sections in the first chapter deal with the special cases when the length of the fourth side exceeds that of the third side by numeric values: 1 - 17. The last section consists of some miscellaneous results.

The Chapter 2 includes the results (in continuation with the preceding chapter) where the fourth side exceeds the third side by values: 18 - 21. The general case when d exceeds c by any integer, say k, is also considered. The last chapter presents the generalizations of Pythagoras theorem to pentagons and includes results of the type $a^2 + b^2 + c^2 + d^2 = e^2$. The cases when e exceeds d by 1, 2 and 3 are included.

The contents are divided into Sections and the discussion within the Sections is presented in the form of Definitions, Theorems, Corollaries, Notes and Examples. The sub-titles within the Sections are numbered in decimal pattern. For instance, the equation number ($c.s.e$) refers to the eth equation in the sth section of Chapter c. When c coincides with the chapter at hand, it is dropped. Adequate references to the results appeared earlier are made in the text avoiding their unnecessary repetition. Double slashes marked at the end of Theorems, Corollaries etc. indicate their completion. For brevity, some set-theoretic notations and symbols are frequently used, e.g. the symbol $\Rightarrow$ means *implies*. A short bibliography of the subject is provided.

Author's long teaching career of more than ***five decades*** at various universities round the globe and research expertise in different fields helped him for lucid presentation of the subject. Author's humble presentation is dedicated to his mathematics teachers and some of the family relatives. I sincerely offer my gratitude to various Universities all over the world especially University of Allahabad, Prayagraj (Allahabad), India; University of Bonn, Bonn (Germany); University of Turin, Turin (Italy); Abdus Salam International Centre for Theoretical Physics, Trieste (Italy); University of Guyana, Georgetown (Guyana);

P.N.G. University of Technology, Lae (Papua New Guinea); Adama Science & Technology University, Adama (Ethiopia); State University of New York, Incheon (South Korea); Divine Word University, Madang (P.N.G.); University of Kurdistan Hewler, Erbil and Lebanese French University, Erbil (Iraq) where I pursued my researches. Sincere thanks are also due to the publisher for their valuable cooperation for bringing the book into limelight in a limited time.

Although proofs are read with utmost care yet an oversight or any discrepancy brought to the notice of the author by the inquisitive readers(s) shall be thankfully acknowledged. The book comes to conclusion at the 150th birth anniversary of the Father of the Nation (Shri Mohandas Karamchand Gandhi, popularly known as *Bapu* – the father to his countrymen).

Lucknow (India): October 2, 2019 Ram Bilas Misra

CHAPTER 1

GENERALIZATIONS OF PYTHAGORAS THEOREM TO QUADRILATERALS

§ 1. Introduction

The celebrated Greek mathematician (more precisely a geometer) Pythagoras was born on the Island of Samos (Greece) during the period about 569 B.C. He left Samos for Egypt around 535 B.C. to study with the priests in temples. He gave a legendary result known as Pythagoras theorem:

Sum of squares of two mutually perpendicular sides in a right triangle equals the square of the hypotenuse, i.e.

$$b^2 + p^2 = h^2,$$

where b, p, h are the lengths of base, perpendicular and hypotenuse of the triangle. In the present chapter an attempt has been made to extend above result by considering the left hand expression as the sum of squares of 3 integers becoming the square of the fourth integer. It deals with quadrilaterals. In terms of symbols we explore the solutions of the quadratic equation

Fig. 1.1

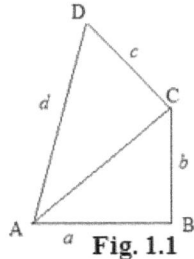

$$a^2 + b^2 + c^2 = d^2. \tag{1.1}$$

Interpreting the result geometrically the integers a, b, c, d satisfying above relation shall represent the lengths of consecutive sides AB, BC, CD and DA respectively of a quadrilateral ABCD composed of 2 right triangles ABC and ACD with right angles at their vertices B and C.

The first eighteen Sections deal with the direct sum of squares of some positive integers making the square of a fourth integer. In the last Section, some special identities involving complicated terms are also discussed.

§ 2. Identities of the type $a^2 + n^2 + b^2 = (b + 1)^2$

Expanding the right hand member, and dropping the common terms one easily derives the value of b in terms of a and n in order to satisfy above identity:

$$b = (a^2 + n^2 - 1)/2, \tag{2.1}$$

where a and n assume some suitable integral values making b integer. For different integral values $a = 1, 2, 3, 4, 5$, etc. above relation yields the following values of b:

a	b	Remark
1	$n^2/2$	b is integer for even n.
2	$(n^2+3)/2 = (n^2+1)/2 + 1$	b is integer for odd n.
3	$(n^2+8)/2 = n^2/2 + 4$	As for $a = 1$.
4	$(n^2+15)/2 = (n^2+1)/2 + 7$	As for $a = 2$.
5	$(n^2+24)/2 = n^2/2 + 12$	As for $a = 1$.
6	$(n^2+35)/2 = (n^2+1)/2 + 17$	As for $a = 2$.
7	$(n^2+48)/2 = n^2/2 + 24$	As for $a = 1$.
8	$(n^2+63)/2 = (n^2+1)/2 + 31$	As for $a = 2$.
9	$(n^2+80)/2 = n^2/2 + 40$	As for $a = 1$.
10	$(n^2+99)/2 = (n^2+1)/2 + 49$	As for $a = 2$.

etc. As such, there exist identities of above type for every integer a. Thus, we have the following theorems.

Theorem 2.1. For $a = 1$, there hold the identities:

n	b	Identity
2	2	$1^2 + 2^2 + 2^2 = 3^2$
4	8	$1^2 + 4^2 + 8^2 = 9^2$
6	18	$1^2 + 6^2 + 18^2 = 19^2$

8	32	$1^2 + 8^2 + 32^2 = 33^2$
10	50	$1^2 + 10^2 + 50^2 = 51^2$
12	72	$1^2 + 12^2 + 72^2 = 73^2$

etc. //

Theorem 2.2. For $a = 2$, here hold the identities:

n	b	Identity	Reference
1	2	$2^2 + 1^2 + 2^2 = 3^2$	Theo. 2.1
3	6	$2^2 + 3^2 + 6^2 = 7^2$	
5	14	$2^2 + 5^2 + 14^2 = 15^2$	
7	26	$2^2 + 7^2 + 26^2 = 27^2$	
9	42	$2^2 + 9^2 + 42^2 = 43^2$	
11	62	$2^2 + 11^2 + 62^2 = 63^2$	
13	86	$2^2 + 13^2 + 86^2 = 87^2$	

etc. //

Theorem 2.3. For $a = 3$, there hold the identities:

n	b	Identity	Reference
2	6	$3^2 + 2^2 + 6^2 = 7^2$	Theo. 2.2
4	12	$3^2 + 4^2 + 12^2 = 13^2$	
6	22	$3^2 + 6^2 + 22^2 = 23^2$	
8	36	$3^2 + 8^2 + 36^2 = 37^2$	
10	54	$3^2 + 10^2 + 54^2 = 55^2$	

12	76	$3^2 + 12^2 + 76^2 = 77^2$	

etc. //

Theorem 2.4. For $a = 4$, there hold the identities:

n	b	Identity	Reference
1	8	$4^2 + 1^2 + 8^2 = 9^2$	Theo. 2.1
3	12	$4^2 + 3^2 + 12^2 = 13^2$	Theo. 2.3
5	20	$4^2 + 5^2 + 20^2 = 21^2$	
7	32	$4^2 + 7^2 + 32^2 = 33^2$	
9	48	$4^2 + 9^2 + 48^2 = 49^2$	
11	68	$4^2 + 11^2 + 68^2 = 69^2$	

etc. //

Theorem 2.5. For $a = 5$, here hold the identities:

n	b	Identity	Reference
2	14	$5^2 + 2^2 + 14^2 = 15^2$	Theo. 2.2
4	20	$5^2 + 4^2 + 20^2 = 21^2$	Theo. 2.4
6	30	$5^2 + 6^2 + 30^2 = 31^2$	
8	44	$5^2 + 8^2 + 44^2 = 45^2$	
10	62	$5^2 + 10^2 + 62^2 = 63^2$	
12	84	$5^2 + 12^2 + 84^2 = 85^2$	

etc. //

Theorem 2.6. For $a = 6$, there hold the identities:

n	b	Identity	Reference
1	18	$6^2 + 1^2 + 18^2 = 19^2$	Theo. 2.1
3	22	$6^2 + 3^2 + 22^2 = 23^2$	Theo. 2.3
5	30	$6^2 + 5^2 + 30^2 = 31^2$	Theo. 2.5
7	42	$6^2 + 7^2 + 42^2 = 43^2$	
9	58	$6^2 + 9^2 + 58^2 = 59^2$	
11	78	$6^2 + 11^2 + 78^2 = 79^2$	

etc. //

Theorem 2.7. For $a = 7$, there hold the identities:

n	b	Identity	Reference
2	26	$7^2 + 2^2 + 26^2 = 27^2$	Theo. 2.2
4	32	$7^2 + 4^2 + 32^2 = 33^2$	Theo. 2.4
6	42	$7^2 + 6^2 + 42^2 = 43^2$	Theo. 2.6
8	56	$7^2 + 8^2 + 56^2 = 57^2$	
10	74	$7^2 + 10^2 + 74^2 = 75^2$	
12	96	$7^2 + 12^2 + 96^2 = 97^2$	

etc. //

Theorem 2.8. For $a = 8$, there hold the identities:

n	b	Identity	Reference
1	32	$8^2 + 1^2 + 32^2 = 33^2$	Theo. 2.1

3	36	$8^2 + 3^2 + 36^2 = 37^2$	Theo. 2.3
5	44	$8^2 + 5^2 + 44^2 = 45^2$	Theo. 2.5
7	56	$8^2 + 7^2 + 56^2 = 57^2$	Theo. 2.7
9	72	$8^2 + 9^2 + 72^2 = 73^2$	
11	92	$8^2 + 11^2 + 92^2 = 93^2$	

etc. //

Theorem 2.9. For $a = 9$, there hold the identities:

n	b	Identity	Reference
2	42	$9^2 + 2^2 + 42^2 = 43^2$	Theo. 2.2
4	48	$9^2 + 4^2 + 48^2 = 49^2$	Theo. 2.4
6	58	$9^2 + 6^2 + 58^2 = 59^2$	Theo. 2.6
8	72	$9^2 + 8^2 + 72^2 = 73^2$	Theo. 2.8
10	90	$9^2 + 10^2 + 90^2 = 91^2$	
12	112	$9^2 + 12^2 + 112^2 = 113^2$	

etc. //

Theorem 2.10. For $a = 10$, there hold the identities:

n	b	Identity	Reference
1	50	$10^2 + 1^2 + 50^2 = 51^2$	Theo. 2.1
3	54	$10^2 + 3^2 + 54^2 = 55^2$	Theo. 2.3
5	62	$10^2 + 5^2 + 62^2 = 63^2$	Theo. 2.5
7	74	$10^2 + 7^2 + 74^2 = 75^2$	Theo. 2.7

9	90	$10^2 + 9^2 + 90^2 = 91^2$	Theo. 2.9
11	110	$10^2 + 11^2 + 110^2 = 111^2$	

etc. //

§ 3. Identities of the type $a^2 + n^2 + b^2 = (b + 2)^2$

Above type of identities require:

$$b = (a^2 + n^2 - 4) / 4 = (a^2 + n^2) / 4 - 1, \qquad (3.1)$$

where a and n assume some suitable integral values making b integer. For $a = 1$, Eq. (3.1) yields $b = (n^2 + 1)/4 - 1$, which cannot assume integral values for any integer n. For instance, when n is even (say $2p$),

$$(n^2 + 1) / 4 = (4p^2 + 1) / 4 = p^2 + 1/4,$$

which is never an integer for any integer p. Similarly, for *odd* values of n (say $2p + 1$),

$$(n^2 + 1) / 4 = (4p^2 + 4p + 2) / 4 = p^2 + p + 1/2,$$

is also not integer. Hence, there exist no such identities for $a = 1$. But, $a = 2 \Rightarrow b = n^2/4$ assuming integral values for any even n. Hence, there exist identities for $a = 2$. In the following, we check for other values of a. Different integral values of $a = 3, 4, 5$, etc. yield the following values of b:

a	b	Remark
3	$(n^2 + 5)/4 = (n^2 + 1)/4 + 1$	As for $a = 1$.
4	$(n^2 + 12)/4 = n^2/4 + 3$	As for $a = 2$.
5	$(n^2 + 21)/4 = (n^2 + 1)/4 + 5$	As for $a = 1$.
6	$(n^2 + 32)/4 = n^2/4 + 8$	As for $a = 2$.
7	$(n^2 + 45)/4 = (n^2 + 1)/4 + 11$	As for $a = 1$.

8	$(n^2 + 60)/4 = n^2/4 + 15$	As for $a = 2$.
9	$(n^2 + 77)/4 = (n^2 + 1)/4 + 19$	As for $a = 1$.
10	$(n^2 + 96)/4 = n^2/4 + 24$	As for $a = 2$.

etc. Conclusively, there exist identities for even values of a, but no identities for odd values of a.

Theorem 3.1. For $a = 2$, there hold the identities:

n	b	Identity	Equivalently	Ref.
2	1	$2^2 + 2^2 + 1^2 = 3^2$		Th. 2.1
4	4	$2^2 + 4^2 + 4^2 = 6^2$	$1^2 + 2^2 + 2^2 = 3^2$	”
6	9	$2^2 + 6^2 + 9^2 = 11^2$		
8	16	$2^2 + 8^2 + 16^2 = 18^2$	$1^2 + 4^2 + 8^2 = 9^2$	Th. 2.1
10	25	$2^2 + 10^2 + 25^2 = 27^2$		
12	36	$2^2 + 12^2 + 36^2 = 38^2$	$1^2 + 6^2 + 18^2 = 19^2$	”

etc. //

Theorem 3.2. For $a = 4$, there hold the following identities:

n	b	Identity	Equivalently	Ref.
2	4	$4^2 + 2^2 + 4^2 = 6^2$	$2^2 + 1^2 + 2^2 = 3^2$	Th. 2.1
4	7	$4^2 + 4^2 + 7^2 = 9^2$		
6	12	$4^2 + 6^2 + 12^2 = 14^2$	$2^2 + 3^2 + 6^2 = 7^2$	Th. 2.2
8	19	$4^2 + 8^2 + 19^2 = 21^2$		
10	28	$4^2 + 10^2 + 28^2 = 30^2$	$2^2 + 5^2 + 14^2 = 15^2$	Th. 2.2

12	39	$4^2 + 12^2 + 39^2 = 41^2$		

etc. //

Theorem 3.3. For $a = 6$, there hold the identities:

n	b	Identity	Equivalently	Ref.
2	9	$6^2 + 2^2 + 9^2 = 11^2$		Th. 3.1
4	12	$6^2 + 4^2 + 12^2 = 14^2$	$3^2 + 2^2 + 6^2 = 7^2$	Th. 2.2
6	17	$6^2 + 6^2 + 17^2 = 19^2$		
8	24	$6^2 + 8^2 + 24^2 = 26^2$	$3^2 + 4^2 + 12^2 = 13^2$	Th. 2.3
10	33	$6^2 + 10^2 + 33^2 = 35^2$		
12	44	$6^2 + 12^2 + 44^2 = 46^2$	$3^2 + 6^2 + 22^2 = 23^2$	Th. 2.3

etc. //

Theorem 3.4. For $a = 8$, there hold the identities:

n	b	Identity	Equivalently	Ref.
2	16	$8^2 + 2^2 + 16^2 = 18^2$	$4^2 + 1^2 + 8^2 = 9^2$	Th. 2.1
4	19	$8^2 + 4^2 + 19^2 = 21^2$		Th. 3.2
6	24	$8^2 + 6^2 + 24^2 = 26^2$	$4^2 + 3^2 + 12^2 = 13^2$	Th. 2.3
8	31	$8^2 + 8^2 + 31^2 = 33^2$		
10	40	$8^2 + 10^2 + 40^2 = 42^2$	$4^2 + 5^2 + 20^2 = 21^2$	Th. 2.4
12	51	$8^2 + 12^2 + 51^2 = 53^2$		

etc. //

Theorem 3.5. For $a = 10$, there hold the identities:

n	b	Identity	Equivalently	Ref.
2	25	$10^2 + 2^2 + 25^2 = 27^2$		Th. 3. 1
4	28	$10^2 + 4^2 + 28^2 = 30^2$	$5^2 + 2^2 + 14^2 = 15^2$	Th. 2.2
6	33	$10^2 + 6^2 + 33^2 = 35^2$		Th. 3.3
8	40	$10^2 + 8^2 + 40^2 = 42^2$	$5^2 + 4^2 + 20^2 = 21^2$	Th. 2.4
10	49	$10^2 + 10^2 + 49^2 = 51^2$		
12	60	$10^2 + 12^2 + 60^2 = 62^2$	$5^2 + 6^2 + 30^2 = 31^2$	Th. 2.5

etc. //

§ 4. Identities of the type $a^2 + n^2 + b^2 = (b + 3)^2$

Above type of identities require:

$$b = (a^2 + n^2 - 9) / 6 = (a^2 + n^2 - 3) / 6 - 1, \qquad (4.1)$$

where a and n assume some suitable integral values making b integer. For different integral values $a = 1, 2, 3, 4, 5$, etc. above relation yields the following values of b:

a	b	Remark
1	$(n^2 - 8)/6 = (n^2 + 4)/6 - 2$	b is not integer for any integer n.
2	$(n^2 - 5)/6 = (n^2 + 1)/6 - 1$	”
3	$n^2 / 6$	b is integer for n = integral multiple of 6.
4	$(n^2 + 7)/6 = (n^2 + 1)/6 + 1$	As for $a = 2$ above.
5	$(n^2 + 16)/6 = (n^2 + 4)/6 + 2$	As for $a = 1$ above.

6	$(n^2 + 27)/6 = (n^2 + 3)/6 + 4$	b is integer for $n = 3, 9, 15, 21, 27$, etc.
7	$(n^2 + 40)/6 = (n^2 + 4)/6 + 6$	As for $a = 1$ above.
8	$(n^2 + 55)/6 = (n^2 + 1)/6 + 9$	As for $a = 2$ above.
9	$(n^2 + 72)/6 = n^2/6 + 12$	As for $a = 3$ above.
10	$(n^2 + 91)/6 = (n^2 + 1)/6 + 15$	As for $a = 2$ above.
11	$(n^2 + 112)/6 = (n^2 + 4)/6 + 18$	As for $a = 1$ above.
12	$(n^2 + 135)/6 = (n^2 + 3)/6 + 22$	As for $a = 6$ above.

etc. Thus, there exist identities whenever a is an integral multiple of 3. Consequently, we have the following theorems.

Theorem 4.1. For $a = 3 \Rightarrow b = n^2 / 6$, there hold the identities:

n	b	Identity	Equivalently	Ref.
6	6	$3^2 + 6^2 + 6^2 = 9^2$	$1^2 + 2^2 + 2^2 = 3^2$	Th. 2.1
12	24	$3^2 + 12^2 + 24^2 = 27^2$	$1^2 + 4^2 + 8^2 = 9^2$	"
18	54	$3^2 + 18^2 + 54^2 = 57^2$	$1^2 + 6^2 + 18^2 = 19^2$	"
24	96	$3^2 + 24^2 + 96^2 = 99^2$	$1^2 + 8^2 + 32^2 = 33^2$	"
30	150	$3^2 + 30^2 + 150^2 = 153^2$	$1^2 + 10^2 + 50^2 = 51^2$	"
36	216	$3^2 + 36^2 + 216^2 = 219^2$	$1^2 + 12^2 + 72^2 = 73^2$	"

etc. //

Theorem 4.2. For $a = 6 \Rightarrow b = (n^2 + 3)/6 + 4$, there hold the identities:

n	b	Identity	Equivalently	Ref.

3	6	$6^2 + 3^2 + 6^2 = 9^2$	$2^2 + 1^2 + 2^2 = 3^2$	Th. 2.1
9	18	$6^2 + 9^2 + 18^2 = 21^2$	$2^2 + 3^2 + 6^2 = 7^2$	Th. 2.2
15	42	$6^2 + 15^2 + 42^2 = 45^2$	$2^2 + 5^2 + 14^2 = 15^2$	"
21	78	$6^2 + 21^2 + 78^2 = 81^2$	$2^2 + 7^2 + 26^2 = 27^2$	"
27	126	$6^2 + 27^2 + 126^2 = 129^2$	$2^2 + 9^2 + 42^2 = 43^2$	"
33	186	$6^2 + 33^2 + 186^2 = 189^2$	$2^2 + 11^2 + 62^2 = 63^2$	"

etc. //

Theorem 4.3. For $a = 9 \Rightarrow b = n^2/6 + 12$, there hold the identities:

n	b	Identity	Equivalently	Ref.
6	18	$9^2 + 6^2 + 18^2 = 21^2$	$3^2 + 2^2 + 6^2 = 7^2$	Th. 2.2
12	36	$9^2 + 12^2 + 36^2 = 39^2$	$3^2 + 4^2 + 12^2 = 13^2$	Th. 2.3
18	66	$9^2 + 18^2 + 66^2 = 69^2$	$3^2 + 6^2 + 22^2 = 23^2$	"
24	108	$9^2 + 24^2 + 108^2 = 111^2$	$3^2 + 8^2 + 36^2 = 37^2$	"
30	162	$9^2 + 30^2 + 162^2 = 165^2$	$3^2 + 10^2 + 54^2 = 55^2$	"
36	228	$9^2 + 36^2 + 228^2 = 231^2$	$3^2 + 12^2 + 76^2 = 77^2$	"

etc. //

Theorem 4.4. For $a = 12 \Rightarrow b = (n^2 + 3)/6 + 22$, there hold the identities:

n	b	Identity	Equivalently	Ref.
3	24	$12^2 + 3^2 + 24^2 = 27^2$	$4^2 + 1^2 + 8^2 = 9^2$	Th. 2.1
9	36	$12^2 + 9^2 + 36^2 = 39^2$	$4^2 + 3^2 + 12^2 = 13^2$	Th. 2.3

15	60	$12^2 + 15^2 + 60^2 = 63^2$	$4^2 + 5^2 + 20^2 = 21^2$	Th. 2.4
21	96	$12^2 + 21^2 + 96^2 = 99^2$	$4^2 + 7^2 + 32^2 = 33^2$	"
27	144	$12^2 + 27^2 + 144^2 = 147^2$	$4^2 + 9^2 + 48^2 = 49^2$	"
33	204	$12^2 + 33^2 + 204^2 = 207^2$	$4^2 + 11^2 + 68^2 = 69^2$	"

etc. //

Theorem 4.5. For $a = 15 \Rightarrow b = n^2/6 + 36$, there hold the identities

n	b	Identity	Equivalently	Ref.
6	42	$15^2 + 6^2 + 42^2 = 45^2$	$5^2 + 2^2 + 14^2 = 15^2$	Th. 2.2
12	60	$15^2 + 12^2 + 60^2 = 63^2$	$5^2 + 4^2 + 20^2 = 21^2$	Th. 2.4
18	90	$15^2 + 18^2 + 90^2 = 93^2$	$5^2 + 6^2 + 30^2 = 31^2$	Th. 2.5
24	132	$15^2 + 24^2 + 132^2 = 135^2$	$5^2 + 8^2 + 44^2 = 45^2$	"
30	186	$15^2 + 30^2 + 186^2 = 189^2$	$5^2 + 10^2 + 62^2 = 63^2$	"
36	252	$15^2 + 36^2 + 252^2 = 255^2$	$5^2 + 12^2 + 84^2 = 85^2$	"

etc. //

Theorem 4.6. For $a = 18 \Rightarrow b = (n^2 + 3)/6 + 52$, there hold the identities:

n	b	Identity	Equivalently	Ref.
3	54	$18^2 + 3^2 + 54^2 = 57^2$	$6^2 + 1^2 + 18^2 = 19^2$	Th. 2.1
9	66	$18^2 + 9^2 + 66^2 = 69^2$	$6^2 + 3^2 + 22^2 = 23^2$	Th. 2.3
15	90	$18^2 + 15^2 + 90^2 = 93^2$	$6^2 + 5^2 + 30^2 = 31^2$	Th. 2.5

21	126	$18^2 + 21^2 + 126^2 = 129^2$	$6^2 + 7^2 + 42^2 = 43^2$	Th. 2.6
27	174	$18^2 + 27^2 + 174^2 = 177^2$	$6^2 + 9^2 + 58^2 = 59^2$	"
33	234	$18^2 + 33^2 + 234^2 = 237^2$	$6^2 + 11^2 + 78^2 = 79^2$	"

etc. //

Theorem 4.7. For $a = 21 \Rightarrow b = n^2/6 + 72$, there hold the identities:

n	b	Identity	Equivalently	Ref.
6	78	$21^2 + 6^2 + 78^2 = 81^2$	$7^2 + 2^2 + 26^2 = 27^2$	Th. 2.2
12	96	$21^2 + 12^2 + 96^2 = 99^2$	$7^2 + 4^2 + 32^2 = 33^2$	Th. 2.4
18	126	$21^2 + 18^2 + 126^2 = 129^2$	$7^2 + 6^2 + 42^2 = 43^2$	Th. 2.6
24	168	$21^2 + 24^2 + 168^2 = 171^2$	$7^2 + 8^2 + 56^2 = 57^2$	Th. 2.7
30	222	$21^2 + 30^2 + 222^2 = 225^2$	$7^2 + 10^2 + 74^2 = 75^2$	"
36	288	$21^2 + 36^2 + 288^2 = 291^2$	$7^2 + 12^2 + 96^2 = 97^2$	"

etc. //

Theorem 4.8. For $a = 24 \Rightarrow b = (n^2 + 3)/6 + 94$, there hold the identities:

n	b	Identity	Equivalently	Ref.
3	96	$24^2 + 3^2 + 96^2 = 99^2$	$8^2 + 1^2 + 32^2 = 33^2$	Th. 2.1
9	108	$24^2 + 9^2 + 108^2 = 111^2$	$8^2 + 3^2 + 36^2 = 37^2$	Th. 2.3
15	132	$24^2 + 15^2 + 132^2 = 135^2$	$8^2 + 5^2 + 44^2 = 45^2$	Th. 2.5
21	168	$24^2 + 21^2 + 168^2 = 171^2$	$8^2 + 7^2 + 56^2 = 57^2$	Th. 2.7

27	216	$24^2 + 27^2 + 216^2 = 219^2$	$8^2 + 9^2 + 72^2 = 73^2$	Th. 2.8
33	276	$24^2 + 33^2 + 276^2 = 279^2$	$8^2 + 11^2 + 92^2 = 93^2$	"

etc. //

Theorem 4.9. For $a = 27 \Rightarrow b = n^2/6 + 120$, there hold the identities:

n	b	Identity	Equivalently	Ref.
6	126	$27^2 + 6^2 + 126^2 = 129^2$	$9^2 + 2^2 + 42^2 = 43^2$	Th. 2.2
12	144	$27^2 + 12^2 + 144^2 = 147^2$	$9^2 + 4^2 + 48^2 = 49^2$	Th. 2.4
18	174	$27^2 + 18^2 + 174^2 = 177^2$	$9^2 + 6^2 + 58^2 = 59^2$	Th. 2.6
24	216	$27^2 + 24^2 + 216^2 = 219^2$	$9^2 + 8^2 + 72^2 = 73^2$	Th. 2.8
30	270	$27^2 + 30^2 + 270^2 = 273^2$	$9^2 + 10^2 + 90^2 = 91^2$	Th. 2.9
36	336	$27^2 + 36^2 + 336^2 = 339^2$	$9^2 + 12^2 + 112^2 = 113^2$	"

etc. //

Theorem 4.10. For $a = 30 \Rightarrow b = (n^2 + 3)/6 + 148$, there hold the identities:

n	b	Identity	Equivalently	Ref.
3	150	$30^2 + 3^2 + 150^2 = 153^2$	$10^2 + 1^2 + 50^2 = 51^2$	Th. 2.1
9	162	$30^2 + 9^2 + 162^2 = 165^2$	$10^2 + 3^2 + 54^2 = 55^2$	Th. 2.3
15	186	$30^2 + 15^2 + 186^2 = 189^2$	$10^2 + 5^2 + 62^2 = 63^2$	Th. 2.5
21	222	$30^2 + 21^2 + 222^2 = 225^2$	$10^2 + 7^2 + 74^2 = 75^2$	Th. 2.7
27	270	$30^2 + 27^2 + 270^2 = 273^2$	$10^2 + 9^2 + 90^2 = 91^2$	Th. 2.9

33	330	$30^2 + 33^2 + 330^2 = 333^2$	$10^2 + 11^2 + 110^2 = 111^2$	Th. 2.10

etc. //

§ 5. Identities of the type $a^2 + n^2 + b^2 = (b + 4)^2$

Above type of identities require:

$$b = (a^2 + n^2 - 16) / 8, \tag{5.1}$$

where a and n assume some suitable integral values making b integer. For different integral values $a = 1, 2, 3, 4, 5$, etc. above relation yields the following values of b:

a	b	Remark
1	$(n^2 - 15)/8 = (n^2 + 1)/8 - 2$	b is not integer for any integer n.
2	$(n^2 - 12)/8 = (n^2 - 4)/8 - 1$	b is integer for $n = 2, 6, 10, 14, 18, 22$, etc.
3	$(n^2 - 7)/8 = (n^2 + 1)/8 - 1$	As for $a = 1$ above.
4	$n^2 / 8$	b is integer if n is an integral multiple of 4.
5	$(n^2 + 9)/8 = (n^2 + 1)/8 + 1$	As for $a = 1$ above.
6	$(n^2 + 20)/8 = (n^2 - 4)/8 + 3$	As for $a = 2$ above.
7	$(n^2 + 33)/8 = (n^2 + 1)/8 + 4$	As for $a = 1$ above.
8	$(n^2 + 48)/8 = n^2/8 + 6$	As for $a = 4$ above.
9	$(n^2 + 65)/8 = (n^2 + 1)/8 + 8$	As for $a = 1$ above.
10	$(n^2 + 84)/8 = (n^2 - 4)/8 + 11$	As for $a = 2$ above.
11	$(n^2 + 105)/8 = (n^2 + 1)/8 + 13$	As for $a = 1$ above.
12	$(n^2 + 128)/8 = n^2/8 + 16$	As for $a = 4$ above.

etc. Thus, there exist identities for every even a. Hence, we have the following theorems.

Theorem 5.1. For $a = 2 \Rightarrow b = (n^2 - 4)/8 - 1$, there hold the following identities for integral values of n, b.

n	b	Identity	Reference
6	3	$2^2 + 6^2 + 3^2 = 7^2$	Theo. 2.2
10	11	$2^2 + 10^2 + 11^2 = 15^2$	
14	23	$2^2 + 14^2 + 23^2 = 27^2$	
18	39	$2^2 + 18^2 + 39^2 = 43^2$	
22	59	$2^2 + 22^2 + 59^2 = 63^2$	
26	83	$2^2 + 26^2 + 83^2 = 87^2$	
30	111	$2^2 + 30^2 + 111^2 = 115^2$	

etc. //

Theorem 5.2. For $a = 4 \Rightarrow b = n^2/8$, there exist identities for n = 4, 8, 12, 16, 20, etc.

n	b	Identity	Equivalently	Ref.
4	2	$4^2 + 4^2 + 2^2 = 6^2$	$2^2 + 2^2 + 1^2 = 3^2$	Th. 2.1
8	8	$4^2 + 8^2 + 8^2 = 12^2$	$1^2 + 2^2 + 2^2 = 3^2$	”
12	18	$4^2 + 12^2 + 18^2 = 22^2$	$2^2 + 6^2 + 9^2 = 11^2$	Th. 3.1
16	32	$4^2 + 16^2 + 32^2 = 36^2$	$1^2 + 4^2 + 8^2 = 9^2$	Th. 2.1
20	50	$4^2 + 20^2 + 50^2 = 54^2$	$2^2 + 10^2 + 25^2 = 27^2$	Th. 3.1
24	72	$4^2 + 24^2 + 72^2 = 76^2$	$1^2 + 6^2 + 18^2 = 19^2$	Th. 2.1

28	98	$4^2 + 28^2 + 98^2 = 102^2$	$2^2 + 14^2 + 49^2 = 51^2$	Th. 3.1
32	128	$4^2 + 32^2 + 128^2 = 132^2$	$1^2 + 8^2 + 32^2 = 33^2$	Th. 2.1

etc. //

Theorem 5.3. For $a = 6 \Rightarrow b = (n^2 - 4)/8 + 3$, there exist identities for $n = 2, 6, 10, 14, 18, 22, 26, 30$, etc.

n	b	Identity	Ref.
2	3	$6^2 + 2^2 + 3^2 = 7^2$	Th. 2.2
6	7	$6^2 + 6^2 + 7^2 = 11^2$	
10	15	$6^2 + 10^2 + 15^2 = 19^2$	
14	27	$6^2 + 14^2 + 27^2 = 31^2$	
18	43	$6^2 + 18^2 + 43^2 = 47^2$	
22	63	$6^2 + 22^2 + 63^2 = 67^2$	
26	87	$6^2 + 26^2 + 87^2 = 91^2$	
30	115	$6^2 + 30^2 + 115^2 = 119^2$	

etc. //

Theorem 5.4. For $a = 8 \Rightarrow b = n^2/8 + 6$, there exist identities for $n = 4, 8, 12, 16, 20$, etc.

n	b	Identity	Equivalently	Ref.
4	8	$8^2 + 4^2 + 8^2 = 12^2$	$2^2 + 1^2 + 2^2 = 3^2$	Th. 2.1
8	14	$8^2 + 8^2 + 14^2 = 18^2$	$4^2 + 4^2 + 7^2 = 9^2$	Th. 3.2
12	24	$8^2 + 12^2 + 24^2 = 28^2$	$2^2 + 3^2 + 6^2 = 7^2$	Th. 2.2
16	38	$8^2 + 16^2 + 38^2 = 42^2$	$4^2 + 8^2 + 19^2 = 21^2$	Th. 3.2

20	56	$8^2 + 20^2 + 56^2 = 60^2$	$2^2 + 5^2 + 14^2 = 15^2$	Th. 2.2
24	78	$8^2 + 24^2 + 78^2 = 82^2$	$4^2 + 12^2 + 39^2 = 41^2$	Th. 3.2

etc. //

Theorem 5.5. For $a = 10 \Rightarrow b = (n^2 - 4)/8 + 11$, there exist identities for $n = 2, 6, 10, 14, 18, 22, 26, 30$, etc.

n	b	Identity	Reference
2	11	$10^2 + 2^2 + 11^2 = 15^2$	Theo. 5.1
6	15	$10^2 + 6^2 + 15^2 = 19^2$	Theo. 5.3
10	23	$10^2 + 10^2 + 23^2 = 27^2$	
14	35	$10^2 + 14^2 + 35^2 = 39^2$	
18	51	$10^2 + 18^2 + 51^2 = 55^2$	
22	71	$10^2 + 22^2 + 71^2 = 75^2$	
26	95	$10^2 + 26^2 + 95^2 = 99^2$	
30	123	$10^2 + 30^2 + 123^2 = 127^2$	

etc. //

Theorem 5.6. For $a = 12 \Rightarrow b = n^2/8 + 16$, there exist identities for $n = 4, 8, 12, 16, 20$, etc.

n	b	Identity	Equivalently	Ref.
4	18	$12^2 + 4^2 + 18^2 = 22^2$	$6^2 + 2^2 + 9^2 = 11^2$	Th. 3.1
8	24	$12^2 + 8^2 + 24^2 = 28^2$	$3^2 + 2^2 + 6^2 = 7^2$	Th. 2.2
12	34	$12^2 + 12^2 + 34^2 = 38^2$	$6^2 + 6^2 + 17^2 = 19^2$	Th. 3.3
16	48	$12^2 + 16^2 + 48^2 = 52^2$	$3^2 + 4^2 + 12^2 = 13^2$	Th. 2.3

20	66	$12^2 + 20^2 + 66^2 = 70^2$	$6^2 + 10^2 + 33^2 = 35^2$	Th. 3.3
24	88	$12^2 + 24^2 + 88^2 = 92^2$	$3^2 + 6^2 + 22^2 = 23^2$	Th. 2.3

etc. //

§ 6. Identities of the type $a^2 + n^2 + b^2 = (b + 5)^2$

Above type of identities require:

$$b = (a^2 + n^2 - 25) / 10, \tag{6.1}$$

where a and n assume some suitable integral values making b integer. For different integral values $a = 1, 2, 3, 4, 5$, etc. above relation yields the following values of b:

a	b	Remark
1	$(n^2 - 24)/10 = (n^2 - 4)/10 - 2$	b is +ve integer for $n = 8, 12, 18, 22, 28, 32$, etc.
2	$(n^2 - 21)/10 = (n^2 - 1)/10 - 2$	b is +ve integer for $n = 9, 11, 19, 21, 29, 31$, etc.
3	$(n^2 - 16)/10 = (n^2 + 4)/10 - 2$	b is +ve integer for $n = 6, 14, 16, 24, 26, 34$, etc.
4	$(n^2 - 9)/10 = (n^2 + 1)/10 - 1$	b is +ve integer for $n = 7, 13, 17, 23, 27, 33$, etc.
5	$n^2 / 10$	b is +ve integer for $n = 10, 20, 30, 40, 50$, etc.
6	$(n^2 + 11)/10 = (n^2 + 1)/10 + 1$	b is +ve integer for $n = 3, 7, 13, 17, 23, 27$, etc.
7	$(n^2 + 24)/10 = (n^2 + 4)/10 + 2$	b is +ve integer for $n = 4, 6, 14, 16, 24, 26$, etc.
8	$(n^2 + 39)/10 = (n^2 - 1)/10 + 4$	b is +ve integer for $n = 1, 9, 11, 19, 21, 29$, etc.
9	$(n^2 + 56)/10 = (n^2 - 4)/10 + 6$	b is +ve integer for $n = 2, 8, 12, 18, 22, 28$, etc.
10	$(n^2 + 75)/10 = (n^2 + 5)/10 + 7$	b is +ve integer for $n = 5, 15, 25, 35, 45$, etc.
11	$(n^2 + 96)/10 = (n^2 - 4)/10 + 10$	As for $a = 9$ above.

12	$(n^2 + 119)/10 = (n^2 - 1)/10 + 12$	As for $a = 8$ above.

etc. Hence, we have the following theorems:

Theorem 6.1. For $a = 1 \Rightarrow b = (n^2 - 4)/10 - 2$, there hold the identities for integral values of n, b.

n	b	Identity	Reference
8	4	$1^2 + 8^2 + 4^2 = 9^2$	Theo. 2.1
12	12	$1^2 + 12^2 + 12^2 = 17^2$	
18	30	$1^2 + 18^2 + 30^2 = 35^2$	
22	46	$1^2 + 22^2 + 46^2 = 51^2$	
28	76	$1^2 + 28^2 + 76^2 = 81^2$	
32	100	$1^2 + 32^2 + 100^2 = 105^2$	

etc. //

Theorem 6.2. For $a = 2 \Rightarrow b = (n^2 - 1)/10 - 2$, there hold the identities for integral values of n, b.

n	b	Identity	Reference
9	6	$2^2 + 9^2 + 6^2 = 11^2$	Theo. 3.1
11	10	$2^2 + 11^2 + 10^2 = 15^2$	Theo. 5.1
19	34	$2^2 + 19^2 + 34^2 = 39^2$	
21	42	$2^2 + 21^2 + 42^2 = 47^2$	
29	82	$2^2 + 29^2 + 82^2 - 87^2$	
31	94	$2^2 + 31^2 + 94^2 = 99^2$	

etc. //

Theorem 6.3. For $a = 3 \Rightarrow b = (n^2 + 4)/10 - 2$, there hold the identities for integral values of n, b.

n	b	Identity	Ref.
6	2	$3^2 + 6^2 + 2^2 = 7^2$	Th. 2.2
14	18	$3^2 + 14^2 + 18^2 = 23^2$	
16	24	$3^2 + 16^2 + 24^2 = 29^2$	
24	56	$3^2 + 24^2 + 56^2 = 61^2$	
26	66	$3^2 + 26^2 + 66^2 = 71^2$	
34	114	$3^2 + 34^2 + 114^2 = 119^2$	

etc. //

Theorem 6.4. For $a = 4 \Rightarrow b = (n^2 + 1)/10 - 1$, there hold the following identities:

n	b	Identity	Ref.
7	4	$4^2 + 7^2 + 4^2 = 9^2$	Th. 3.2
13	16	$4^2 + 13^2 + 16^2 = 21^2$	
17	28	$4^2 + 17^2 + 28^2 = 33^2$	
23	52	$4^2 + 23^2 + 52^2 = 57^2$	
27	72	$4^2 + 27^2 + 72^2 = 77^2$	
33	108	$4^2 + 33^2 + 108^2 = 113^2$	

etc. //

Theorem 6.5. When $a = 5 \Rightarrow b = n^2 / 10$, there hold the identities for positive integral multiples of $n = 10$.

n	b	Identity	Equivalently	Ref.
10	10	$5^2 + 10^2 + 10^2 = 15^2$	$1^2 + 2^2 + 2^2 = 3^2$	Th. 2.1
20	40	$5^2 + 20^2 + 40^2 = 45^2$	$1^2 + 4^2 + 8^2 = 9^2$	”
30	90	$5^2 + 30^2 + 90^2 = 95^2$	$1^2 + 6^2 + 18^2 = 19^2$	”
40	160	$5^2 + 40^2 + 160^2 = 165^2$	$1^2 + 8^2 + 32^2 = 33^2$	”
50	250	$5^2 + 50^2 + 250^2 = 255^2$	$1^2 + 10^2 + 50^2 = 51^2$	”
60	360	$5^2 + 60^2 + 360^2 = 365^2$	$1^2 + 12^2 + 72^2 = 73^2$	”

etc. //

Theorem 6.6. For $a = 6 \Rightarrow b = (n^2 + 1)/10 + 1$, there hold the following identities:

n	b	Identity	Ref.
3	2	$6^2 + 3^2 + 2^2 = 7^2$	Th. 2.2
7	6	$6^2 + 7^2 + 6^2 = 11^2$	Th. 5.3
13	18	$6^2 + 13^2 + 18^2 = 23^2$	
17	30	$6^2 + 17^2 + 30^2 = 35^2$	
23	54	$6^2 + 23^2 + 54^2 = 59^2$	
27	74	$6^2 + 27^2 + 74^2 = 79^2$	

etc. //

Theorem 6.7. For $a = 7 \Rightarrow b = (n^2 + 4)/10 + 2$, there hold the identities:

n	b	Identity	Ref.
4	4	$7^2 + 4^2 + 4^2 = 9^2$	Th. 3.2
6	6	$7^2 + 6^2 + 6^2 = 11^2$	Th. 5.3
14	22	$7^2 + 14^2 + 22^2 = 27^2$	
16	28	$7^2 + 16^2 + 28^2 = 33^2$	
24	60	$7^2 + 24^2 + 60^2 = 65^2$	
26	70	$7^2 + 26^2 + 70^2 = 75^2$	

etc. //

Theorem 6.8. For $a = 8 \Rightarrow b = (n^2 - 1)/10 + 4$, there hold the identities:

n	b	Identity	Ref.
1	4	$8^2 + 1^2 + 4^2 = 9^2$	Th. 2.1
9	12	$8^2 + 9^2 + 12^2 = 17^2$	
11	16	$8^2 + 11^2 + 16^2 = 21^2$	
19	40	$8^2 + 19^2 + 40^2 = 45^2$	
21	48	$8^2 + 21^2 + 48^2 = 53^2$	
29	88	$8^2 + 29^2 + 88^2 = 93^2$	

etc. //

Theorem 6.9. For $a = 9 \Rightarrow b = (n^2 - 4)/10 + 6$, there hold the identities for integral values of n, b.

n	b	Identity	Ref.
2	6	$9^2 + 2^2 + 6^2 = 11^2$	Th. 3.1
8	12	$9^2 + 8^2 + 12^2 = 17^2$	Th. 6.8
12	20	$9^2 + 12^2 + 20^2 = 25^2$	
18	38	$9^2 + 18^2 + 38^2 = 43^2$	
22	54	$9^2 + 22^2 + 54^2 = 59^2$	
28	84	$9^2 + 28^2 + 84^2 = 89^2$	

etc. //

Theorem 6.10. For $a = 10 \Rightarrow b = (n^2 + 5)/10 + 7$, there hold the identities for integral values of n, b.

n	b	Identity	Equivalently	Ref.
5	10	$10^2 + 5^2 + 10^2 = 15^2$	$2^2 + 1^2 + 2^2 = 3^2$	Th. 2.1
15	30	$10^2 + 15^2 + 30^2 = 35^2$	$2^2 + 3^2 + 6^2 = 7^2$	Th. 2.2
25	70	$10^2 + 25^2 + 70^2 = 75^2$	$2^2 + 5^2 + 14^2 = 15^2$	”
35	130	$10^2 + 35^2 + 130^2 = 135^2$	$2^2 + 7^2 + 26^2 = 27^2$	”
45	210	$10^2 + 45^2 + 210^2 = 215^2$	$2^2 + 9^2 + 42^2 = 43^2$	”

etc. //

§ 7. Identities of the type $a^2 + n^2 + b^2 = (b + 6)^2$

Above type of identities require:

$$b = (a^2 + n^2 - 36) / 12, \tag{7.1}$$

where a and n assume some suitable integral values making b integer. For $a = 1, 2, 3, 4, 5$, etc. above relation yields values of b:

a	b	Remark
1	$(n^2 - 35)/12 = (n^2 + 1)/12 - 3$	b is not an integer for any integer n.
2	$(n^2 - 32)/12 = (n^2 + 4)/12 - 3$	”
3	$(n^2 - 27)/12 = (n^2 - 3)/12 - 2$	”
4	$(n^2 - 20)/12 = (n^2 + 4)/12 - 2$	”
5	$(n^2 - 11)/12 = (n^2 + 1)/12 - 1$	”
6	$n^2/ 12$	b is integer when n is an integral multiple of 6.
7	$(n^2 + 13)/12 = (n^2 + 1)/12 + 1$	b is not an integer for any integer n.
8	$(n^2 + 28)/12 = (n^2 + 4)/12 + 2$	”
9	$(n^2 + 45)/12 = (n^2 - 3)/12 + 4$	”
10	$(n^2 + 64)/12 = (n^2 + 4)/12 + 5$	”
11	$(n^2 + 85)/12 = (n^2 + 1)/12 + 7$	”
12	$(n^2 + 108)/12 = n^2 /12 + 9$	As for $a = 6$ above.

etc. Similarly, $a = 13$ to 17 also do not yield any integral values of b for any integer n. But, $a = 18 \Rightarrow b = (n^2 + 288)/12 = n^2 /12 + 24$, which in analogy with $a = 6$, takes integral values for any integral multiple of $n = 6$. Thus, there hold the following theorems:

Theorem 7.1. For $a = 6 \Rightarrow b = n^2/12$, there hold the identities for integral values of n, b.

n	b	Identity	Equivalently	Ref.
6	3	$6^2 + 6^2 + 3^2 = 9^2$	$2^2 + 2^2 + 1^2 = 3^2$	Th. 2.1

12	12	$6^2 + 12^2 + 12^2 = 18^2$	$1^2 + 2^2 + 2^2 = 3^2$	”
18	27	$6^2 + 18^2 + 27^2 = 33^2$	$2^2 + 6^2 + 9^2 = 11^2$	Th. 3.1
24	48	$6^2 + 24^2 + 48^2 = 54^2$	$1^2 + 4^2 + 8^2 = 9^2$	Th. 2.1
30	75	$6^2 + 30^2 + 75^2 = 81^2$	$2^2 + 10^2 + 25^2 = 27^2$	Th. 3.1
36	108	$6^2 + 36^2 + 108^2 = 114^2$	$1^2 + 6^2 + 18^2 = 19^2$	Th. 2.1

etc. //

Theorem 7.2. For $a = 12 \Rightarrow b = n^2/12 + 9$, there hold the identities for integral values of n, b.

n	b	Identity	Equivalently	Ref.
6	12	$12^2 + 6^2 + 12^2 = 18^2$	$2^2 + 1^2 + 2^2 = 3^2$	Th. 2.1
12	21	$12^2 + 12^2 + 21^2 = 27^2$	$4^2 + 4^2 + 7^2 = 9^2$	Th. 3.2
18	36	$12^2 + 18^2 + 36^2 = 42^2$	$2^2 + 3^2 + 6^2 = 7^2$	Th. 2.2
24	57	$12^2 + 24^2 + 57^2 = 63^2$	$4^2 + 8^2 + 19^2 = 21^2$	Th. 3.2
30	84	$12^2 + 30^2 + 84^2 = 90^2$	$2^2 + 5^2 + 14^2 = 15^2$	Th. 2.2
36	117	$12^2 + 36^2 + 117^2 = 123^2$	$4^2 + 12^2 + 39^2 = 41^2$	Th. 3.2

etc. //

Theorem 7.3. For $a = 18 \Rightarrow b = n^2/12 + 24$, there hold the identities for integral values of n, b.

n	b	Identity	Equivalently	Ref.
6	27	$18^2 + 6^2 + 27^2 = 33^2$	$6^2 + 2^2 + 9^2 = 11^2$	Th. 3.1
12	36	$18^2 + 12^2 + 36^2 = 42^2$	$3^2 + 2^2 + 6^2 = 7^2$	Th. 2.2

18	51	$18^2 + 18^2 + 51^2 = 57^2$	$6^2 + 6^2 + 17^2 = 19^2$	Th. 3.3
24	72	$18^2 + 24^2 + 72^2 = 78^2$	$3^2 + 4^2 + 12^2 = 13^2$	Th. 2.3
30	99	$18^2 + 30^2 + 99^2 = 105^2$	$6^2 + 10^2 + 33^2 = 35^2$	Th. 3.3
36	132	$18^2 + 36^2 + 132^2 = 138^2$	$3^2 + 6^2 + 22^2 = 23^2$	Th. 2.3

etc. //

Note 7.1. Conclusively, there exist such identities for every integral multiple of $a = 6$.

§ 8. Identities of the type $a^2 + n^2 + b^2 = (b + 7)^2$

Above type of identities require:

$$b = (a^2 + n^2 - 49) / 14, \tag{8.1}$$

where a and n assume some suitable integral values making b integer. For $a = 1, 2, 3, 4, 5$, etc. above relation yields values of b:

a	b	Remark
1	$(n^2 - 48)/14 = (n^2 - 6)/14 - 3$	b is not an integer for any integer n.
2	$(n^2 - 45)/14 = (n^2 - 3)/14 - 3$	”
3	$(n^2 - 40)/14 = (n^2 + 2)/14 - 3$	”
4	$(n^2 - 33)/14 = (n^2 - 5)/14 - 2$	”
5	$(n^2 - 24)/14 = (n^2 + 4)/14 - 2$	”
6	$(n^2 - 13)/14 = (n^2 + 1)/14 - 1$	”
7	$n^2/ 14$	b is integer for n = an integral multiple of 14.
8	$(n^2 + 15)/14 = (n^2 + 1)/14 + 1$	As for $a = 6$ above.

9	$(n^2 + 32)/14 = (n^2 + 4)/14 + 2$	As for $a = 5$ above.
10	$(n^2 + 51)/14 = (n^2 - 5)/14 + 4$	As for $a = 4$ above.
11	$(n^2 + 72)/14 = (n^2 + 2)/14 + 5$	As for $a = 3$ above.
12	$(n^2 + 95)/14 = (n^2 - 3)/14 + 7$	As for $a = 2$ above.
13	$(n^2 + 120)/14 = (n^2 - 6)/14 + 9$	As for $a = 1$ above.
14	$(n^2 + 147)/14 = (n^2 + 7)/14 + 10$	b is integer for $n = 7, 21, 35, 49, 63, 77$, etc.

Therefore, there hold such identities only when a is an integral multiple of 7 giving the following theorems:

Theorem 8.1. For $a = 7 \Rightarrow b = n^2/14$, there hold the following identities for integral values of n, b.

n	b	Identity	Equivalently	Ref.
14	14	$7^2 + 14^2 + 14^2 = 21^2$	$1^2 + 2^2 + 2^2 = 3^2$	Th. 2.1
28	56	$7^2 + 28^2 + 56^2 = 63^2$	$1^2 + 4^2 + 8^2 = 9^2$	”
42	126	$7^2 + 42^2 + 126^2 = 133^2$	$1^2 + 6^2 + 18^2 = 19^2$	”
56	224	$7^2 + 56^2 + 224^2 = 231^2$	$1^2 + 8^2 + 32^2 = 33^2$	”
70	350	$7^2 + 70^2 + 350^2 = 357^2$	$1^2 + 10^2 + 50^2 = 51^2$	”
84	504	$7^2 + 84^2 + 504^2 = 511^2$	$1^2 + 12^2 + 72^2 = 73^2$	”

etc. //

Theorem 8.2. For $a = 14 \Rightarrow b = (n^2 + 7)/14 + 10$, there hold the identities for integral values of n, b.

n	b	Identity	Equivalently	Ref.
7	14	$14^2 + 7^2 + 14^2 = 21^2$	$2^2 + 1^2 + 2^2 = 3^2$	Th. 2.1
21	42	$14^2 + 21^2 + 42^2 = 49^2$	$2^2 + 3^2 + 6^2 = 7^2$	Th. 2.2
35	98	$14^2 + 35^2 + 98^2 = 105^2$	$2^2 + 5^2 + 14^2 = 15^2$	”
49	182	$14^2 + 49^2 + 182^2 = 189^2$	$2^2 + 7^2 + 26^2 = 27^2$	”
63	294	$14^2 + 63^2 + 294^2 = 301^2$	$2^2 + 9^2 + 42^2 = 43^2$	”
77	434	$14^2 + 77^2 + 434^2 = 441^2$	$2^2 + 11^2 + 62^2 = 63^2$	”

etc. //

Similarly, there do not exist any such identities for a = 15 to 20. However, for $a = 21$, Eq. (8.1) yields

$$b = (n^2 + 392) / 14 = n^2 / 14 + 28, \qquad (8.2)$$

which, in analogy with Theo. 8.1, take integral values when n is an integral multiple of 14. Thus, there holds the theorem:

Theorem 8.3. For $a = 21$, by Eq. (8.2), we have the identities for integral values of n, b.

n	b	Identity	Equivalently	Ref.
14	42	$21^2 + 14^2 + 42^2 = 49^2$	$3^2 + 2^2 + 6^2 = 7^2$	Th. 2.2
28	84	$21^2 + 28^2 + 84^2 = 91^2$	$3^2 + 4^2 + 12^2 = 13^2$	Th. 2.3
42	154	$21^2 + 42^2 + 154^2 = 161^2$	$3^2 + 6^2 + 22^2 = 23^2$	”
56	252	$21^2 + 56^2 + 252^2 = 259^2$	$3^2 + 8^2 + 36^2 = 37^2$	”

70	378	$21^2 + 70^2 + 378^2 = 385^2$	$3^2 + 10^2 + 54^2 = 55^2$	”

etc. //

For $a = 22$ to 27 also there exist no identities but $a = 28$ yields

$$b = (n^2 + 735) / 14 = (n^2 + 7) / 14 + 52, \tag{8.3}$$

which, in analogy with Theo. 8.2, take integral values for $n = 7, 21, 35, 49$, etc. Thus, there holds the theorem:

Theorem 8.4. For $a = 28$, by Eq. (8.3), there hold the following identities for integral values of n, b.

n	b	Identity	Equivalently	Ref.
7	56	$28^2 + 7^2 + 56^2 = 63^2$	$4^2 + 1^2 + 8^2 = 9^2$	Th. 2.1
21	84	$28^2 + 21^2 + 84^2 = 91^2$	$4^2 + 3^2 + 12^2 = 13^2$	Th. 2.3
35	140	$28^2 + 35^2 + 140^2 = 147^2$	$4^2 + 5^2 + 20^2 = 21^2$	Th. 2.4
49	224	$28^2 + 49^2 + 224^2 = 231^2$	$4^2 + 7^2 + 32^2 = 33^2$	”
63	336	$28^2 + 63^2 + 336^2 = 343^2$	$4^2 + 9^2 + 48^2 = 49^2$	”

etc. //

§ 9. Identities of the type $a^2 + n^2 + b^2 = (b + 8)^2$

Above type of identities require:

$$b = (a^2 + n^2 - 64) / 16, \tag{9.1}$$

where a and n assume some suitable integral values making b integer. For $a - 1, 2, 3, 4, 5$, etc. above relation yields values of b:

a	b	Remark
1	$(n^2 - 63)/16 = (n^2 + 1)/16 - 4$	b is not an integer for any integer n.

2	$(n^2 - 60)/16 = (n^2 + 4)/16 - 4$	"
3	$(n^2 - 55)/16 = (n^2 - 7)/16 - 3$	"
4	$(n^2 - 48)/16 = n^2/16 - 3$	b is integer when n is an integral multiple of 4
5	$(n^2 - 39)/16 = (n^2 - 7)/16 - 2$	As for $a = 3$ above.
6	$(n^2 - 28)/16 = (n^2 + 4)/16 - 2$	As for $a = 2$ above.
7	$(n^2 - 15)/16 = (n^2 + 1)/16 - 1$	As for $a = 1$ above.
8	$n^2/16$	As for $a = 4$ above.
9	$(n^2 + 17)/16 = (n^2 + 1)/16 + 1$	As for $a = 1$ above.
10	$(n^2 + 36)/16 = (n^2 + 4)/16 + 2$	As for $a = 2$ above.
11	$(n^2 + 57)/16 = (n^2 - 7)/16 + 4$	As for $a = 3$ above.
12	$(n^2 + 80)/16 = n^2/16 + 5$	As for $a = 4$ above.

Thus, there exist such identities only when a is an integral multiple of 4 giving rise to the following theorems:

Theorem 9.1. For $a = 4 \Rightarrow b = n^2/16 - 3$, there hold the identities for integral values of n, b.

n	b	Identity	Equivalently	Ref.
8	1	$4^2 + 8^2 + 1^2 = 9^2$		Th. 2.1
12	6	$4^2 + 12^2 + 6^2 = 14^2$	$2^2 + 6^2 + 3^2 = 7^2$	Th. 2.2
16	13	$4^2 + 16^2 + 13^2 = 21^2$		Th. 6.4
20	22	$4^2 + 20^2 + 22^2 = 30^2$	$2^2 + 10^2 + 11^2 = 15^2$	Th. 5.1
24	33	$4^2 + 24^2 + 33^2 = 41^2$		

28	46	$4^2 + 28^2 + 46^2 = 54^2$	$2^2 + 14^2 + 23^2 = 27^2$	Th. 5.1
32	61	$4^2 + 32^2 + 61^2 = 69^2$		

etc. //

Theorem 9.2. For $a = 8 \Rightarrow b = n^2/16$, there hold the identities for integral values of n, b.

n	b	Identity	Equivalently	Ref.
4	1	$8^2 + 4^2 + 1^2 = 9^2$		Th. 2.1
8	4	$8^2 + 8^2 + 4^2 = 12^2$	$2^2 + 2^2 + 1^2 = 3^2$	”
12	9	$8^2 + 12^2 + 9^2 = 17^2$		Th. 6.8
16	16	$8^2 + 16^2 + 16^2 = 24^2$	$1^2 + 2^2 + 2^2 = 3^2$	Th. 2.1
20	25	$8^2 + 20^2 + 25^2 = 33^2$		
24	36	$8^2 + 24^2 + 36^2 = 44^2$	$2^2 + 6^2 + 9^2 = 11^2$	Th. 3.1
28	49	$8^2 + 28^2 + 49^2 = 57^2$		
32	64	$8^2 + 32^2 + 64^2 = 72^2$	$1^2 + 4^2 + 8^2 = 9^2$	Th. 2.1

etc. //

Theorem 9.3. For $a = 12 \Rightarrow b = n^2/16 + 5$, there hold the identities for integral values of n, b.

n	b	Identity	Equivalently	Ref.
4	6	$12^2 + 4^2 + 6^2 = 14^2$	$6^2 + 2^2 + 3^2 = 7^2$	Th. 2.2
8	9	$12^2 + 8^2 + 9^2 = 17^2$		Th. 6.8
12	14	$12^2 + 12^2 + 14^2 = 22^2$	$6^2 + 6^2 + 7^2 = 11^2$	Th. 5.3
16	21	$12^2 + 16^2 + 21^2 = 29^2$		

20	30	$12^2 + 20^2 + 30^2 = 38^2$	$6^2 + 10^2 + 15^2 = 19^2$	Th. 5.3
24	41	$12^2 + 24^2 + 41^2 = 49^2$		
28	54	$12^2 + 28^2 + 54^2 = 62^2$	$6^2 + 14^2 + 27^2 = 31^2$	Th. 5.3
32	69	$12^2 + 32^2 + 69^2 = 77^2$		

etc. //

§ 10. Identities of the type $a^2 + n^2 + b^2 = (b + 9)^2$

Above type of identities require:

$$b = (a^2 + n^2 - 81) / 18, \tag{10.1}$$

where a and n assume some suitable integral values making b integer. For $a = 1, 2, 3, 4, 5$, etc. above relation yields values of b:

a	b	Remark
1	$(n^2 - 80)/18 = (n^2 - 8)/18 - 4$	b is not an integer for any integer n.
2	$(n^2 - 77)/18 = (n^2 - 5)/18 - 4$	"
3	$(n^2 - 72)/18 = n^2/18 - 4$	b is integer when n is an integral multiple of 6.
4	$(n^2 - 65)/18 = (n^2 + 7)/18 - 4$	b is not an integer for any integer n.
5	$(n^2 - 56)/18 = (n^2 - 2)/18 - 3$	"
6	$(n^2 - 45)/18 = (n^2 - 9)/18 - 2$	b is integer for $n = 3, 9, 15, 21, 27, 33, 39$, etc.
7	$(n^2 - 32)/18 = (n^2 + 4)/18 - 2$	b is not an integer for any integer n.
8	$(n^2 - 17)/18 = (n^2 + 1)/18 - 1$	"
9	$n^2/18$	As for $a = 3$ above.
10	$(n^2 + 19)/18 = (n^2 + 1)/18 + 1$	As for $a = 8$ above.

11	$(n^2 + 40)/18 = (n^2 + 4)/18 + 2$	As for $a = 7$ above.
12	$(n^2 + 63)/18 = (n^2 - 9)/18 + 4$	As for $a = 6$ above.

etc. Hence, there exist such identities only when n is an integral multiple of 3 giving rise to the following theorems:

Theorem 10.1. For $a = 3 \Rightarrow b = n^2/18 - 4$, there hold the identities for integral values of n, b.

n	b	Identity	Ref.
12	4	$3^2 + 12^2 + 4^2 = 13^2$	Th. 2.3
18	14	$3^2 + 18^2 + 14^2 = 23^2$	Th. 6.3
24	28	$3^2 + 24^2 + 28^2 = 37^2$	
30	46	$3^2 + 30^2 + 46^2 = 55^2$	
36	68	$3^2 + 36^2 + 68^2 = 77^2$	
42	94	$3^2 + 42^2 + 94^2 = 103^2$	
48	124	$3^2 + 48^2 + 124^2 = 133^2$	

etc. //

Theorem 10.2. For $a = 6 \Rightarrow b = (n^2 - 9)/18 - 2$, there hold the following identities for integral values of n, b.

n	b	Identity	Ref.
9	2	$6^2 + 9^2 + 2^2 = 11^2$	Th. 3.1
15	10	$6^2 + 15^2 + 10^2 = 19^2$	Th. 5.3
21	22	$6^2 + 21^2 + 22^2 = 31^2$	
27	38	$6^2 + 27^2 + 38^2 = 47^2$	

33	58	$6^2 + 33^2 + 58^2 = 67^2$	
39	82	$6^2 + 39^2 + 82^2 = 91^2$	
45	110	$6^2 + 45^2 + 110^2 = 119^2$	

etc. //

Theorem 10.3. For $a = 9 \Rightarrow b = n^2/18$, there hold the identities for integral values of n, b.

n	b	Identity	Equivalently	Ref.
6	2	$9^2 + 6^2 + 2^2 = 11^2$		Th. 3.1
12	8	$9^2 + 12^2 + 8^2 = 17^2$		Th. 6.8
18	18	$9^2 + 18^2 + 18^2 = 27^2$	$1^2 + 2^2 + 2^2 = 3^2$	Th. 2.1
24	32	$9^2 + 24^2 + 32^2 = 41^2$		
30	50	$9^2 + 30^2 + 50^2 = 59^2$		
36	72	$9^2 + 36^2 + 72^2 = 81^2$	$1^2 + 4^2 + 8^2 = 9^2$	Th. 2.1
42	98	$9^2 + 42^2 + 98^2 = 107^2$		
48	128	$9^2 + 48^2 + 128^2 = 137^2$		

etc. //

Theorem 10.4. For $a = 12 \Rightarrow b = (n^2 - 9)/18 + 4$, there hold the identities for integral values of n, b.

n	b	Identity	Ref.
3	4	$12^2 + 3^2 + 4^2 = 13^2$	Th. 2.3
9	8	$12^2 + 9^2 + 8^2 = 17^2$	Th. 6.8
15	16	$12^2 + 15^2 + 16^2 = 25^2$	

21	28	$12^2 + 21^2 + 28^2 = 37^2$	
27	44	$12^2 + 27^2 + 44^2 = 53^2$	
33	64	$12^2 + 33^2 + 64^2 = 73^2$	
39	88	$12^2 + 39^2 + 88^2 = 97^2$	
45	116	$12^2 + 45^2 + 116^2 = 125^2$	

etc. //

§ 11. Identities of the type $a^2 + n^2 + b^2 = (b + 10)^2$

Above type of identities require:

$$b = (a^2 + n^2 - 100)/20 = (a^2 + n^2)/20 - 5, \qquad (11.1)$$

where a and n assume some suitable integral values making b integer. For positive integral values of a above relation yields the following values of b:

a	b	Remarks
1	$(n^2 + 1) / 20 - 5$	b is not an integer for any integer n.
2	$(n^2 + 4) / 20 - 5$	b is an integer for $n = 4, 6, 14, 16, 24, 26$, etc. $\Rightarrow$ identities.
3	$(n^2 + 9) / 20 - 5$	b is not an integer for any integer n.
4	$(n^2 - 4) / 20 - 4$	b is an integer for $n = 2, 8, 12, 18, 22, 28$, etc. $\Rightarrow$ identities.
5	$(n^2 + 5) / 20 - 4$	b is not an integer for any integer n.
6	$(n^2 - 4) / 20 - 3$	As for $a = 4 \Rightarrow$ identities.
7	$(n^2 + 9)/20 - 3$	As for $a = 3 \Rightarrow$ no identities.
8	$(n^2 + 4)/20 - 2$	As for $a = 2 \Rightarrow$ identities.
9	$(n^2 + 1)/20 - 1$	As for $a = 1 \Rightarrow$ no identities.

10	$n^2/20$	b is integer when n is integral multiple of 10 $\Rightarrow$ identities.

etc. Conclusively, there exist identities for even values of a giving rise to the following theorems:

Theorem 11.1. For $a = 2 \Rightarrow b = (n^2 + 4)/20 - 5$, there hold the identities for positive integral values of n, b.

n	b	Identity	Equivalently	Ref.
14	5	$2^2 + 14^2 + 5^2 = 15^2$		Th. 2.2
16	8	$2^2 + 16^2 + 8^2 = 18^2$	$1^2 + 8^2 + 4^2 = 9^2$	Th. 2.1
24	24	$2^2 + 24^2 + 24^2 = 34^2$	$1^2 + 12^2 + 12^2 = 17^2$	Th. 6.1
26	29	$2^2 + 26^2 + 29^2 = 39^2$		
34	53	$2^2 + 34^2 + 53^2 = 63^2$		
36	60	$2^2 + 36^2 + 60^2 = 70^2$	$1^2 + 18^2 + 30^2 = 35^2$	Th. 6.1

etc. //

Theorem 11.2. For $a = 4 \Rightarrow b = (n^2 - 4) / 20 - 4$, there hold the following identities for positive integral values of n, b.

n	b	Identity	Equivalently	Ref.
12	3	$4^2 + 12^2 + 3^2 = 13^2$		Th. 2.3
18	12	$4^2 + 18^2 + 12^2 = 22^2$	$2^2 + 9^2 + 6^2 = 11^2$	Th. 3.1
22	20	$4^2 + 22^2 + 20^2 = 30^2$	$2^2 + 11^2 + 10^2 = 15^2$	Th. 5.1
28	35	$4^2 + 28^2 + 35^2 = 45^2$		
32	47	$4^2 + 32^2 + 47^2 = 57^2$		
38	68	$4^2 + 38^2 + 68^2 = 78^2$	$2^2 + 19^2 + 34^2 = 39^2$	Th. 6.2

etc. //

Theorem 11.3. For $a = 6 \Rightarrow b = (n^2 - 4) / 20 - 3$, there hold the following identities for positive integral values of n, b.

n	b	Identity	Equivalently	Ref.
12	4	$6^2 + 12^2 + 4^2 = 14^2$	$3^2 + 6^2 + 2^2 = 7^2$	Th. 2.2
18	13	$6^2 + 18^2 + 13^2 = 23^2$		
22	21	$6^2 + 22^2 + 21^2 = 31^2$		
28	36	$6^2 + 28^2 + 36^2 = 46^2$	$3^2 + 14^2 + 18^2 = 23^2$	Th. 6.3
32	48	$6^2 + 32^2 + 48^2 = 58^2$	$3^2 + 16^2 + 24^2 = 29^2$	"
38	69	$6^2 + 38^2 + 69^2 = 79^2$		

etc. //

Theorem 11.4. For $a = 8 \Rightarrow b = (n^2 + 4)/20 - 2$, there hold the identities for positive integral values of n, b.

n	b	Identity	Equivalently	Ref.
14	8	$8^2 + 14^2 + 8^2 = 18^2$	$4^2 + 7^2 + 4^2 = 9^2$	Th. 3.2
16	11	$8^2 + 16^2 + 11^2 = 21^2$		
24	27	$8^2 + 24^2 + 27^2 = 37^2$		
26	32	$8^2 + 26^2 + 32^2 = 42^2$	$4^2 + 13^2 + 16^2 = 21^2$	Th. 6.4
34	56	$8^2 + 34^2 + 56^2 = 66^2$	$4^2 + 17^2 + 28^2 = 33^2$	"
36	63	$8^2 + 36^2 + 63^2 = 73^2$		

etc. //

Theorem 11.5. For $a = 10 \Rightarrow b = n^2/20$, there hold the identities for integral values of n, b.

n	b	Identity	Equivalently	Ref.
10	5	$10^2 + 10^2 + 5^2 = 15^2$	$2^2 + 2^2 + 1^2 = 3^2$	Th. 2.1
20	20	$10^2 + 20^2 + 20^2 = 30^2$	$1^2 + 2^2 + 2^2 = 3^2$	”
30	45	$10^2 + 30^2 + 45^2 = 55^2$	$2^2 + 6^2 + 9^2 = 11^2$	Th. 3.1
40	80	$10^2 + 40^2 + 80^2 = 90^2$	$1^2 + 4^2 + 8^2 = 9^2$	Th. 2.1
50	125	$10^2 + 50^2 + 125^2 = 135^2$	$2^2 + 10^2 + 25^2 = 27^2$	Th. 3.1
60	180	$10^2 + 60^2 + 180^2 = 190^2$	$1^2 + 6^2 + 18^2 = 19^2$	Th. 2.1

etc. //

§ 12. Identities of the type $a^2 + n^2 + b^2 = (b + 11)^2$

Above type of identities require:

$$b = (a^2 + n^2 - 121) / 22, \tag{12.1}$$

where a and n assume some suitable integral values making b integer. For positive integral values of a above relation yields the following values of b:

a	b
1	$(n^2 - 120) / 22 = (n^2 - 10) / 22 - 5$
2	$(n^2 - 117) / 22 = (n^2 - 7) / 22 - 5$
3	$(n^2 - 112) / 22 = (n^2 - 2) / 22 - 5$
4	$(n^2 - 105) / 22 = (n^2 + 5) / 22 - 5$
5	$(n^2 - 96) / 22 = (n^2 - 8) / 22 - 4$

6	$(n^2 - 85) / 22 = (n^2 + 3) / 22 - 4$
7	$(n^2 - 72) / 22 = (n^2 - 6) / 22 - 3$
8	$(n^2 - 57) / 22 = (n^2 + 9) / 22 - 3$
9	$(n^2 - 40) / 22 = (n^2 + 4) / 22 - 2$
10	$(n^2 - 21) / 22 = (n^2 + 1) / 22 - 1$
11	$n^2 / 22$

Case 12.1. For odd values of n, $n^2 - 2$, $n^2 - 6$, $n^2 - 8$, $n^2 - 10$, $n^2 + 4$ are all odd; hence not divisible by 22. Eventually, there exist no integral values of b. On the other hand, $n^2 + 1$, $n^2 + 3$, $n^2 + 5$, $n^2 + 9$, $n^2 - 7$ are even but still not divisible by 22.

Case 12.2. For even values of n, $n^2 - 2$, $n^2 - 6$, $n^2 - 8$, $n^2 - 10$, $n^2 + 4$ are all even but not divisible by 22; while $n^2 + 1$, $n^2 + 3$, $n^2 + 5$, $n^2 + 9$, $n^2 - 7$ being odd are also not divisible by 22.

Thus, there exist no integral values of b for $a = 1$ to 10 implying no identities of above type. On the other hand, $n^2/22$ takes integral values when n is an integral multiple of 22 giving rise to the following theorem:

Theorem 12.1. For $a = 11 \Rightarrow b = n^2/22$, there hold the identities for integral values of n, b.

n	b	Identity	Equivalently	Ref.
22	22	$11^2 + 22^2 + 22^2 = 33^2$	$1^2 + 2^2 + 2^2 = 3^2$	Th. 2.1
44	88	$11^2 + 44^2 + 88^2 = 99^2$	$1^2 + 4^2 + 8^2 = 9^2$	”
66	198	$11^2 + 66^2 + 198^2 = 209^2$	$1^2 + 6^2 + 18^2 = 19^2$	”
88	352	$11^2 + 88^2 + 352^2 = 363^2$	$1^2 + 8^2 + 32^2 = 33^2$	”
110	550	$11^2 + 110^2 + 550^2 = 561^2$	$1^2 + 10^2 + 50^2 = 51^2$	”

etc. //

Similarly, for $a = 12$ to 21, we get

$$b = (n^2 + 23) / 22 = (n^2 + 1) / 22 + 1, \dots,$$

$$(n^2 + 320) / 22 = (n^2 - 10) / 22 + 15;$$

which also do not take integral values for any integer n. Hence, there exist no such identities for these values of a. However, $a = 22$ yields integral values of

$$b = (n^2 + 363) / 22 = (n^2 + 11) / 22 + 16 \qquad (12.2)$$

taken for $n = 11, 33, 55, 77, 99$, etc. giving rise to the theorem:

Theorem 12.2. For $a = 22$ and Eq. (12.2), there hold the identities for integral values of n, b.

n	b	Identity	Equivalently	Ref.
11	22	$22^2 + 11^2 + 22^2 = 33^2$	$2^2 + 1^2 + 2^2 = 3^2$	Th. 2.1
33	66	$22^2 + 33^2 + 66^2 = 77^2$	$2^2 + 3^2 + 6^2 = 7^2$	Th. 2.2
55	154	$22^2 + 55^2 + 154^2 = 165^2$	$2^2 + 5^2 + 14^2 = 15^2$	”
77	286	$22^2 + 77^2 + 286^2 = 297^2$	$2^2 + 7^2 + 26^2 = 27^2$	”
99	462	$22^2 + 99^2 + 462^2 = 473^2$	$2^2 + 9^2 + 42^2 = 43^2$	”

etc. //

For $a = 33$, $b = n^2/22 + 44$ takes integral values when n is any integral multiple of 22 giving rise to the theorem:

Theorem 12.3. For $a = 33$, there hold the identities for integral values of n, b.

n	b	Identity	Equivalently	Ref.
22	66	$33^2 + 22^2 + 66^2 = 77^2$	$3^2 + 2^2 + 6^2 = 7^2$	Th. 2.2
44	132	$33^2 + 44^2 + 132^2 = 143^2$	$3^2 + 4^2 + 12^2 = 13^2$	Th. 2.3
66	242	$33^2 + 66^2 + 242^2 = 253^2$	$3^2 + 6^2 + 22^2 = 23^2$	”
88	396	$33^2 + 88^2 + 396^2 = 407^2$	$3^2 + 8^2 + 36^2 = 37^2$	”
110	594	$33^2 + 110^2 + 594^2 = 605^2$	$3^2 + 10^2 + 54^2 = 55^2$	”

etc. //

For $a = 44$, $b = (n^2 + 1815)/22 = (n^2 + 11)/22 + 82$, which in analogy with Theo. 12.2, take integral values for $n = 11, 33, 55, 77, 99$, etc. Thus, we have the:

Theorem 12.4. For $a = 44$, there hold the identities for integral values of n, b.

n	b	Identity	Equivalently	Ref.
11	88	$44^2 + 11^2 + 88^2 = 99^2$	$4^2 + 1^2 + 8^2 = 9^2$	Th. 2.1
33	132	$44^2 + 33^2 + 132^2 = 143^2$	$4^2 + 3^2 + 12^2 = 13^2$	Th. 2.3
55	220	$44^2 + 55^2 + 220^2 - 231^2$	$4^2 + 5^2 + 20^2 = 21^2$	Th. 2.4
77	352	$44^2 + 77^2 + 352^2 = 363^2$	$4^2 + 7^2 + 32^2 = 33^2$	”
99	528	$44^2 + 99^2 + 528^2 = 539^2$	$4^2 + 9^2 + 48^2 = 49^2$	”

etc. //

For $a = 55$, $b = n^2/22 + 132$ takes integral values when n is any integral multiple of 22. Thus, we have the:

Theorem 12.5. For $a = 55$, there hold the identities for integral values of n, b.

n	b	Identity	Equivalently	Ref.
22	154	$55^2 + 22^2 + 154^2 = 165^2$	$5^2 + 2^2 + 14^2 = 15^2$	Th. 2.2
44	220	$55^2 + 44^2 + 220^2 = 231^2$	$5^2 + 4^2 + 20^2 = 21^2$	Th. 2.4
66	330	$55^2 + 66^2 + 330^2 = 341^2$	$5^2 + 6^2 + 30^2 = 31^2$	Th. 2.5
88	484	$55^2 + 88^2 + 484^2 = 495^2$	$5^2 + 8^2 + 44^2 = 45^2$	"
110	682	$55^2 + 110^2 + 682^2 = 693^2$	$5^2 + 10^2 + 62^2 = 63^2$	"

etc. //

For $a = 66$, $b = (n^2 + 11) /22 + 192$, which in analogy with Theo. 12.2, take integral values for $n = 11, 33, 55, 77, 99$, etc. Thus, we have the:

Theorem 12.6. For $a = 66$, there hold the following identities for integral values of n, b.

n	b	Identity	Equivalently	Ref.
11	198	$66^2 + 11^2 + 198^2 = 209^2$	$6^2 + 1^2 + 18^2 = 19^2$	Th. 2.1
33	242	$66^2 + 33^2 + 242^2 = 253^2$	$6^2 + 3^2 + 22^2 = 23^2$	Th. 2.3
55	330	$66^2 + 55^2 + 330^2 = 341^2$	$6^2 + 5^2 + 30^2 = 31^2$	Th. 2.5
77	462	$66^2 + 77^2 + 462^2 = 473^2$	$6^2 + 7^2 + 42^2 = 43^2$	Th. 2.6
99	638	$66^2 + 99^2 + 638^2 = 649^2$	$6^2 + 9^2 + 58^2 = 59^2$	"

etc. //

§ 13. Identities of the type $a^2 + n^2 + b^2 = (b + 12)^2$

Above type of identities require:

$$b = (a^2 + n^2 - 144) / 24 = (a^2 + n^2) / 24 - 6, \qquad (13.1)$$

where a and n assume some suitable integral values making b integer. For positive integral values of a, above relation yields the following values of b:

a	b	Remark
1	$(n^2 - 143)/24 = (n^2 + 1)/24 - 6$	b is not an integer for any integer n.
2	$(n^2 - 140)/24 = (n^2 + 4)/24 - 6$	”
3	$(n^2 - 135)/24 = (n^2 + 9)/24 - 6$	”
4	$(n^2 - 128)/24 = (n^2 - 8)/24 - 5$	”
5	$(n^2 - 119)/24 = (n^2 + 1)/24 - 5$	”
6	$(n^2 - 108)/24 = (n^2 + 12)/24 - 5$	b is integer when n = 18,30,42,54,66, etc.
7	$(n^2 - 95)/24 = (n^2 + 1)/24 - 4$	b is not an integer for any integer n.
8	$(n^2 - 80)/24 = (n^2 - 8)/24 - 3$	”
9	$(n^2 - 63)/24 = (n^2 + 9)/24 - 3$	”
10	$(n^2 - 44)/24 = (n^2 + 4)/24 - 2$	”
11	$(n^2 - 23)/24 = (n^2 + 1)/24 - 1$	”
12	$n^2 / 24$	b is integer for n = integral multiple of 12 $\Rightarrow$ identities
18	$(n^2 + 180)/24 = (n^2 + 12)/24 + 7$	As for $a = 6$.
24	$n^2/24 + 18$	As for $a = 12$.
36	$n^2/24 + 48$	”

etc. Thus, we note that there exist such identities only when a is an integral multiple of 6 giving rise to the following theorems:

Theorem 13.1. For $a = 6 \Rightarrow b = (n^2 + 12)/24 - 5$, there hold the identities for integral values of n, b.

n	b	Identity	Equivalently	Ref.
18	9	$6^2 + 18^2 + 9^2 = 21^2$	$2^2 + 6^2 + 3^2 = 7^2$	Th. 2.2
30	33	$6^2 + 30^2 + 33^2 = 45^2$	$2^2 + 10^2 + 11^2 = 15^2$	Th. 5.1
42	69	$6^2 + 42^2 + 69^2 = 81^2$	$2^2 + 14^2 + 23^2 = 27^2$	”
54	117	$6^2 + 54^2 + 117^2 = 129^2$	$2^2 + 18^2 + 39^2 = 43^2$	”
66	177	$6^2 + 66^2 + 177^2 = 189^2$	$2^2 + 22^2 + 59^2 = 63^2$	”

etc. //

Theorem 13.2. For $a = 12 \Rightarrow b = n^2/24$, there hold the following identities for integral values of n, b.

n	b	Identity	Equivalently	Ref.
12	6	$12^2 + 12^2 + 6^2 = 18^2$	$2^2 + 2^2 + 1^2 = 3^2$	Th. 2.1
24	24	$12^2 + 24^2 + 24^2 = 36^2$	$1^2 + 2^2 + 2^2 = 3^2$	”
36	54	$12^2 + 36^2 + 54^2 = 66^2$	$2^2 + 6^2 + 9^2 = 11^2$	Th. 3.1
48	96	$12^2 + 48^2 + 96^2 = 108^2$	$1^2 + 4^2 + 8^2 = 9^2$	Th. 2.1
60	150	$12^2 + 60^2 + 150^2 = 162^2$	$2^2 + 10^2 + 25^2 = 27^2$	Th. 3.1

etc. //

Theorem 13.3. For $a = 18 \Rightarrow b = (n^2 + 12)/24 + 7$, there hold the identities for integral values of n, b.

n	b	Identity	Equivalently	Ref.
6	9	$18^2 + 6^2 + 9^2 = 21^2$	$6^2 + 2^2 + 3^2 = 7^2$	Th. 2.2
18	21	$18^2 + 18^2 + 21^2 = 33^2$	$6^2 + 6^2 + 7^2 = 11^2$	Th. 5.3
30	45	$18^2 + 30^2 + 45^2 = 57^2$	$6^2 + 10^2 + 15^2 = 19^2$	”
42	81	$18^2 + 42^2 + 81^2 = 93^2$	$6^2 + 14^2 + 27^2 = 31^2$	”
54	129	$18^2 + 54^2 + 129^2 = 141^2$	$6^2 + 18^2 + 43^2 = 47^2$	”
66	189	$18^2 + 66^2 + 189^2 = 201^2$	$6^2 + 22^2 + 63^2 = 67^2$	”

etc. //

Theorem 13.4. For $a = 24 \Rightarrow b = n^2/24 + 18$, there hold the following identities for integral values of n, b.

n	b	Identity	Equivalently	Ref.
12	24	$24^2 + 12^2 + 24^2 = 36^2$	$2^2 + 1^2 + 2^2 = 3^2$	Th. 2.1
24	42	$24^2 + 24^2 + 42^2 = 54^2$	$4^2 + 4^2 + 7^2 = 9^2$	Th. 3.2
36	72	$24^2 + 36^2 + 72^2 = 84^2$	$2^2 + 3^2 + 6^2 = 7^2$	Th. 2.2
48	114	$24^2 + 48^2 + 114^2 = 126^2$	$4^2 + 8^2 + 19^2 = 21^2$	Th. 3.2
60	168	$24^2 + 60^2 + 168^2 = 180^2$	$2^2 + 5^2 + 14^2 = 15^2$	Th. 2.2

etc. //

Theorem 13.5. For $a = 36 \Rightarrow b = n^2/24 + 48$, there hold the following identities for integral values of n, b.

n	b	Identity	Equivalently	Ref.
12	54	$36^2 + 12^2 + 54^2 = 66^2$	$6^2 + 2^2 + 9^2 = 11^2$	Th. 3.1
24	72	$36^2 + 24^2 + 72^2 = 84^2$	$3^2 + 2^2 + 6^2 = 7^2$	Th. 2.2
36	102	$36^2 + 36^2 + 102^2 = 114^2$	$6^2 + 6^2 + 17^2 = 19^2$	Th. 3.3
48	144	$36^2 + 48^2 + 144^2 = 156^2$	$3^2 + 4^2 + 12^2 = 13^2$	Th. 2.3
60	198	$36^2 + 60^2 + 198^2 = 210^2$	$6^2 + 10^2 + 33^2 = 35^2$	Th. 3.3

etc. //

§ 14. Identities of the type $a^2 + n^2 + b^2 = (b + 13)^2$

Above type of identities require:

$$b = (a^2 + n^2 - 169) / 26, \tag{14.1}$$

where a and n assume some suitable integral values making b integer. For positive integral values of a above relation yields the following values of b:

a	b	Remark
1	$(n^2 - 168)/26 = (n^2 - 12)/26 - 6$	b is +ve integer for $n = 18$, 34, 44, 60, 70, etc.
2	$(n^2 - 165)/26 = (n^2 - 9)/26 - 6$	b is +ve integer for $n = 23$, 29, 49, 55, 75, 81, etc.
3	$(n^2 - 160)/26 = (n^2 - 4)/26 - 6$	b is +ve integer for $n = 24$, 28, 50, 54, 76, etc.
4	$(n^2 - 153)/26 = (n^2 + 3)/26 - 6$	b is +ve integer for $n = 19$, 33, 45, 59, 71, etc.
5	$(n^2 - 144)/26 = (n^2 + 12)/26 - 6$	b is +ve integer for $n = 14$, 38, 40, 64, 66, etc.
6	$(n^2 - 133)/26 = (n^2 - 3)/26 - 5$	b is +ve integer for $n = 17$, 35, 43, 61, 69, etc.

7	$(n^2 - 120)/26 = (n^2 + 10)/26 - 5$	b is +ve integer for n = 22, 30, 48, 56, 74, 82, etc.
8	$(n^2 - 105)/26 = (n^2 - 1)/26 - 4$	b is +ve integer for n = 25, 27, 51, 53, 77, 79, etc.
9	$(n^2 - 88)/26 = (n^2 - 10)/26 - 3$	b is integer when n = 20, 32, 46, 58, 72, etc.
10	$(n^2 - 69)/26 = (n^2 + 9)/26 - 3$	b is +ve integer for n = 11, 15, 37, 41, 63, 67, etc.
11	$(n^2 - 48)/26 = (n^2 + 4)/26 - 2$	b is +ve integer for n = 10, 16, 36, 42, 62, 68, etc.
12	$(n^2 - 25)/26 = (n^2 + 1)/26 - 1$	b is +ve integer for n = 21, 31, 47, 57, 73, 83, etc.
13	$n^2 / 26$	b is integer when n is an integral multiple of 26

etc.

Thus, we have the:

Theorem 14.1. For $a = 1$, there hold the identities for integral values of n, b.

n	b	Identity	Ref.
18	6	$1^2 + 18^2 + 6^2 = 19^2$	Th. 2.1
34	38	$1^2 + 34^2 + 38^2 = 51^2$	
44	68	$1^2 + 44^2 + 68^2 = 81^2$	
60	132	$1^2 + 60^2 + 132^2 = 145^2$	
70	182	$1^2 + 70^2 + 182^2 = 195^2$	

etc. //

Theorem 14.2. For $a = 2$, there hold the identities for integral values of n, b.

n	b	Identity	Ref.
23	14	$2^2 + 23^2 + 14^2 = 27^2$	Th.5.1

29	26	$2^2 + 29^2 + 26^2 = 39^2$	
49	86	$2^2 + 49^2 + 86^2 = 99^2$	
55	110	$2^2 + 55^2 + 110^2 = 123^2$	
75	210	$2^2 + 75^2 + 210^2 = 223^2$	
81	246	$2^2 + 81^2 + 246^2 = 259^2$	

etc. //

Theorem 14.3. For $a = 3$, there hold the identities for integral values of n, b.

n	b	Identity	Ref.
24	16	$3^2 + 24^2 + 16^2 = 29^2$	Th. 6.3
28	24	$3^2 + 28^2 + 24^2 = 37^2$	Th.10.1
50	90	$3^2 + 50^2 + 90^2 = 103^2$	
54	106	$3^2 + 54^2 + 106^2 = 119^2$	
76	216	$3^2 + 76^2 + 216^2 = 229^2$	
80	240	$3^2 + 80^2 + 240^2 = 253^2$	

etc. //

Theorem 14.4. For $a = 4$, there hold the identities for integral values of n, b.

n	b	Identity	Ref.
19	8	$4^2 + 19^2 + 8^2 = 21^2$	Th. 3.2
33	36	$4^2 + 33^2 + 36^2 = 49^2$	
45	72	$4^2 + 45^2 + 72^2 = 85^2$	

59	128	$4^2 + 59^2 + 128^2 = 141^2$	
71	188	$4^2 + 71^2 + 188^2 = 201^2$	

etc. //

Theorem 14.5. For $a = 5$, there hold the identities for integral values of n, b.

n	b	Identity	Ref.
14	2	$5^2 + 14^2 + 2^2 = 15^2$	Th. 2.2
38	50	$5^2 + 38^2 + 50^2 = 63^2$	
40	56	$5^2 + 40^2 + 56^2 = 69^2$	
64	152	$5^2 + 64^2 + 152^2 = 165^2$	
66	162	$5^2 + 66^2 + 162^2 = 175^2$	

etc. //

Theorem 14.6. For $a = 6$, there hold the identities for integral values of n, b.

n	b	Identity	Ref.
17	6	$6^2 + 17^2 + 6^2 = 19^2$	Th. 3.3
35	42	$6^2 + 35^2 + 42^2 = 55^2$	
43	66	$6^2 + 43^2 + 66^2 = 79^2$	
61	138	$6^2 + 61^2 + 138^2 = 151^2$	
69	178	$6^2 + 69^2 + 178^2 = 191^2$	

etc. //

Theorem 14.7. For $a = 7$, there hold the identities for integral values of n, b.

n	b	Identity	Ref.
22	14	$7^2 + 22^2 + 14^2 = 27^2$	Th. 6.7
30	30	$7^2 + 30^2 + 30^2 = 43^2$	
48	84	$7^2 + 48^2 + 84^2 = 97^2$	
56	116	$7^2 + 56^2 + 116^2 = 129^2$	
74	206	$7^2 + 74^2 + 206^2 = 219^2$	
82	254	$7^2 + 82^2 + 254^2 = 267^2$	

etc. //

Theorem 14.8. For $a = 8$, there hold the identities for integral values of n, b.

n	b	Identity	Ref.
25	20	$8^2 + 25^2 + 20^2 = 33^2$	Th. 9.2
27	24	$8^2 + 27^2 + 24^2 = 37^2$	
51	96	$8^2 + 51^2 + 96^2 = 109^2$	
53	104	$8^2 + 53^2 + 104^2 = 117^2$	
77	224	$8^2 + 77^2 + 224^2 = 237^2$	
79	236	$8^2 + 79^2 + 236^2 = 249^2$	

etc. //

Theorem 14.9. For $a = 9$, there hold the identities for integral values of n, b.

n	b	Identity	Ref.
20	12	$9^2 + 20^2 + 12^2 = 25^2$	Th. 6.9
32	36	$9^2 + 32^2 + 36^2 = 49^2$	
46	78	$9^2 + 46^2 + 78^2 = 91^2$	
58	126	$9^2 + 58^2 + 126^2 = 139^2$	
72	196	$9^2 + 72^2 + 196^2 = 209^2$	

etc. //

Theorem 14.10. For $a = 10$, there hold the identities for integral values of n, b.

n	b	Identity	Ref.
11	2	$10^2 + 11^2 + 2^2 = 15^2$	Th. 5.1
15	6	$10^2 + 15^2 + 6^2 = 19^2$	Th. 5.3
37	50	$10^2 + 37^2 + 50^2 = 63^2$	
41	62	$10^2 + 41^2 + 62^2 = 75^2$	
63	150	$10^2 + 63^2 + 150^2 = 163^2$	
67	170	$10^2 + 67^2 + 170^2 = 183^2$	

etc. //

Theorem 14.11. For $a = 11$, there hold the identities for integral values of n, b.

n	b	Identity	Ref.
10	2	$11^2 + 10^2 + 2^2 = 15^2$	Th. 5.1

16	8	$11^2 + 16^2 + 8^2 = 21^2$	Th. 6.8
36	48	$11^2 + 36^2 + 48^2 = 61^2$	
42	66	$11^2 + 42^2 + 66^2 = 79^2$	
62	146	$11^2 + 62^2 + 146^2 = 159^2$	
68	176	$11^2 + 68^2 + 176^2 = 189^2$	

etc. //

Theorem 14.12. For $a = 12$, there hold the identities for integral values of n, b.

n	b	Identity	Ref.
21	16	$12^2 + 21^2 + 16^2 = 29^2$	Th. 9.3
31	36	$12^2 + 31^2 + 36^2 = 49^2$	
47	84	$12^2 + 47^2 + 84^2 = 97^2$	
57	124	$12^2 + 57^2 + 124^2 = 137^2$	
73	204	$12^2 + 73^2 + 204^2 = 217^2$	
83	264	$12^2 + 83^2 + 264^2 = 277^2$	

etc. //

Theorem 14.13. For $a = 13$, there hold the identities for integral values of n, b.

n	b	Identity	Equivalently	Ref.
26	26	$13^2 + 26^2 + 26^2 = 39^2$	$1^2 + 2^2 + 2^2 = 3^2$	Th. 2.1
52	104	$13^2 + 52^2 + 104^2 = 117^2$	$1^2 + 4^2 + 8^2 = 9^2$	”

78	234	$13^2 + 78^2 + 234^2 = 247^2$	$1^2 + 6^2 + 18^2 = 19^2$	”
104	416	$13^2 + 104^2 + 416^2 = 429^2$	$1^2 + 8^2 + 32^2 = 33^2$	”
130	650	$13^2 + 130^2 + 650^2 = 663^2$	$1^2 + 10^2 + 50^2 = 51^2$	”

etc. //

§ 15. Identities of the type $a^2 + n^2 + b^2 = (b + 14)^2$

Above type of identities require:

$$b = (a^2 + n^2 - 196) / 28 = (a^2 + n^2) / 28 - 7, \qquad (15.1)$$

where a and n assume some suitable integral values making b integer. For positive integral values of a above relation yields the following values of b:

a	b
1	$(n^2 + 1) / 28 - 7$
2	$(n^2 + 4) / 28 - 7$
3	$(n^2 + 9) / 28 - 7$
4	$(n^2 - 180)/28 = (n^2 - 12)/28 - 6$
5	$(n^2 - 171)/28 = (n^2 - 3)/28 - 6$
6	$(n^2 - 160)/28 = (n^2 + 8)/28 - 6$
7	$(n^2 - 147)/28 = (n^2 - 7)/28 - 5$
8	$(n^2 - 132)/28 = (n^2 + 8)/28 - 5$
9	$(n^2 - 115)/28 = (n^2 - 3)/28 - 4$
10	$(n^2 - 96)/28 = (n^2 - 12)/28 - 3$
11	$(n^2 - 75)/28 = (n^2 + 9)/28 - 3$

12	$(n^2 - 52)/28 = (n^2 + 4)/28 - 2$
13	$(n^2 - 27)/28 = (n^2 + 1)/28 - 1$
14	$n^2 / 28$

Note 15.1. None of the values of $a = 1$ to 13 yield integral values of b. However, $a = 14$ makes b integer when n is an integral multiple of 14 yielding the:

Theorem 15.1. For $a = 14$, there hold the identities for integral values of n, b.

n	b	Identity	Equivalently	Ref.
14	7	$14^2 + 14^2 + 7^2 = 21^2$	$2^2 + 2^2 + 1^2 = 3^2$	Th. 2.1
28	28	$14^2 + 28^2 + 28^2 = 42^2$	$1^2 + 2^2 + 2^2 = 3^2$	”
42	63	$14^2 + 42^2 + 63^2 = 77^2$	$2^2 + 6^2 + 9^2 = 11^2$	Th. 3.1
56	112	$14^2 + 56^2 + 112^2 = 126^2$	$1^2 + 4^2 + 8^2 = 9^2$	Th. 2.1
70	175	$14^2 + 70^2 + 175^2 = 189^2$	$2^2 + 10^2 + 25^2 = 27^2$	Th. 3.1

etc. //

Similarly, $a = 15$ to 27 also do not yield any integral values of b for any integer n. In the following we check the situation when a is an integral multiple of 14.

a	b	Remark
28	$n^2 / 28 + 21$	b is +ve integer for n = integral multiple of 14
42	$n^2/28 + 56$	”
56	$n^2/28 + 105$	”
70	$n^2/28 + 168$	”

etc. yielding the following theorems:

Theorem 15.2. For $a = 28 \Rightarrow b = n^2 / 28 + 21$, there hold the identities for integral values of n, b.

n	b	Identity	Equivalently	Ref.
14	28	$28^2 + 14^2 + 28^2 = 42^2$	$2^2 + 1^2 + 2^2 = 3^2$	Th. 2.1
28	49	$28^2 + 28^2 + 49^2 = 63^2$	$4^2 + 4^2 + 7^2 = 9^2$	Th. 3.2
42	84	$28^2 + 42^2 + 84^2 = 98^2$	$2^2 + 3^2 + 6^2 = 7^2$	Th. 2.2
56	133	$28^2 + 56^2 + 133^2 = 147^2$	$4^2 + 8^2 + 19^2 = 21^2$	Th. 3.2
70	196	$28^2 + 70^2 + 196^2 = 210^2$	$2^2 + 5^2 + 14^2 = 15^2$	Th. 2.2

etc. //

Theorem 15.3. For $a = 42 \Rightarrow b = n^2 / 28 + 56$, there hold the identities for integral values of n, b.

n	b	Identity	Equivalently	Ref.
14	63	$42^2 + 14^2 + 63^2 = 77^2$	$6^2 + 2^2 + 9^2 = 11^2$	Th. 3.1
28	84	$42^2 + 28^2 + 84^2 = 98^2$	$3^2 + 2^2 + 6^2 = 7^2$	Th. 2.2
42	119	$42^2 + 42^2 + 119^2 = 133^2$	$6^2 + 6^2 + 17^2 = 19^2$	Th. 3.3
56	168	$42^2 + 56^2 + 168^2 = 182^2$	$3^2 + 4^2 + 12^2 = 13^2$	Th. 2.3
70	231	$42^2 + 70^2 + 231^2 = 245^2$	$6^2 + 10^2 + 33^2 = 35^2$	Th. 3.3

etc. //

Theorem 15.4. For $a = 56 \Rightarrow b = n^2 / 28 + 105$, there hold the identities for integral values of n, b.

n	b	Identity	Equivalently	Ref.
14	112	$56^2 + 14^2 + 112^2 = 126^2$	$4^2 + 1^2 + 8^2 = 9^2$	Th. 2.1
28	133	$56^2 + 28^2 + 133^2 = 147^2$	$8^2 + 4^2 + 19^2 = 21^2$	Th. 3.2
42	168	$56^2 + 42^2 + 168^2 = 182^2$	$4^2 + 3^2 + 12^2 = 13^2$	Th. 2.3
56	217	$56^2 + 56^2 + 217^2 = 231^2$	$8^2 + 8^2 + 31^2 = 33^2$	Th. 3.4
70	280	$56^2 + 70^2 + 280^2 = 294^2$	$4^2 + 5^2 + 20^2 = 21^2$	Th. 2.4

etc. //

Theorem 15.5. For $a = 70 \Rightarrow b = n^2 / 28 + 168$, there hold the identities for integral values of n, b.

n	b	Identity	Equivalently	Ref.
14	175	$70^2 + 14^2 + 175^2 = 189^2$	$10^2 + 2^2 + 25^2 = 27^2$	Th. 3.1
28	196	$70^2 + 28^2 + 196^2 = 210^2$	$5^2 + 2^2 + 14^2 = 15^2$	Th. 2.2
42	231	$70^2 + 42^2 + 231^2 = 245^2$	$10^2 + 6^2 + 33^2 = 35^2$	Th. 3.3
56	280	$70^2 + 56^2 + 280^2 = 294^2$	$5^2 + 4^2 + 20^2 = 21^2$	Th. 2.4
70	343	$70^2 + 70^2 + 343^2 = 357^2$	$10^2 + 10^2 + 49^2 = 51^2$	Th. 3.5

etc. //

§ 16. Identities of the type $a^2 + n^2 + b^2 = (b + 15)^2$

Above type of identities require:

$$b = (a^2 + n^2 - 225)/30 = (a^2 + n^2 - 15)/30 - 7, \qquad (16.1)$$

where a and n assume some suitable integral values making b integer. For positive integral values of a above relation yields the following values of b:

a	b	Remark
1	$(n^2 - 14) / 30 - 7$	b is not integer for any integer n.
2	$(n^2 - 11) / 30 - 7$	"
3	$(n^2 - 6) / 30 - 7$	b is +ve integer for $n = 24$, 36, 54, 66, 84, etc.
4	$(n^2 + 1) / 30 - 7$	b is not integer for any integer n.
5	$(n^2 + 10) / 30 - 7$	"
6	$(n^2 - 189)/30 = (n^2 - 9)/30 - 6$	b is +ve integer for $n = 27$, 33, 57, 63, 87, 93, etc.
7	$(n^2 - 176)/30 = (n^2 + 4)/30 - 6$	b is not integer for any integer n.
8	$(n^2 - 161)/30 = (n^2 - 11)/30 - 5$	"
9	$(n^2 - 144)/30 = (n^2 + 6)/30 - 5$	b is +ve integer for $n = 18$, 42, 48, 72, 78, 102, etc.
10	$(n^2 - 125)/30 = (n^2 - 5)/30 - 4$	b is not integer for any integer n.
11	$(n^2 - 104)/30 = (n^2 - 14)/30 - 3$	"
12	$(n^2 - 81)/30 = (n^2 + 9)/30 - 3$	b is +ve integer for $n = 21$, 39, 51, 69, 81, 99, etc.
13	$(n^2 - 56)/30 = (n^2 + 4)/30 - 2$	b is not integer for any integer n.
14	$(n^2 - 29)/30 = (n^2 + 1)/30 - 1$	"
15	$n^2 / 30$	b is +ve integer for n = integral multiple of 30.
18	$(n^2 + 99)/30 = (n^2 + 9)/30 + 3$	b is +ve integer for $n = 9$, 21, 39, 51, 69, 81, etc.
21	$(n^2 + 216)/30 = (n^2 + 6)/30 + 7$	b is +ve integer for $n = 12$, 18, 42, 48, 72, 78, etc.

etc. Conclusively, b takes integral values whenever a is an integral multiple of 3. Hence, there hold the following theorems:

Theorem 16.1. For $a = 3$, there hold the identities for integral values of n, b.

n	b	Identity	Equivalently	Ref.
24	12	$3^2 + 24^2 + 12^2 = 27^2$	$1^2 + 8^2 + 4^2 = 9^2$	Th. 2.1
36	36	$3^2 + 36^2 + 36^2 = 51^2$	$1^2 + 12^2 + 12^2 = 17^2$	Th. 6.1
54	90	$3^2 + 54^2 + 90^2 = 105^2$	$1^2 + 18^2 + 30^2 = 35^2$	”
66	138	$3^2 + 66^2 + 138^2 = 153^2$	$1^2 + 22^2 + 46^2 = 51^2$	”
84	228	$3^2 + 84^2 + 228^2 = 243^2$	$1^2 + 28^2 + 76^2 = 81^2$	”

etc. //

Theorem 16.2. For $a = 6$, there hold the identities for integral values of n, b.

n	b	Identity	Equivalently	Ref.
27	18	$6^2 + 27^2 + 18^2 = 33^2$	$2^2 + 9^2 + 6^2 = 11^2$	Th. 3.1
33	30	$6^2 + 33^2 + 30^2 = 45^2$	$2^2 + 11^2 + 10^2 = 15^2$	Th. 5.1
57	102	$6^2 + 57^2 + 102^2 = 117^2$	$2^2 + 19^2 + 34^2 = 39^2$	Th. 6.2
63	126	$6^2 + 63^2 + 126^2 = 141^2$	$2^2 + 21^2 + 42^2 = 47^2$	”
87	246	$6^2 + 87^2 + 246^2 = 261^2$	$2^2 + 29^2 + 82^2 = 87^2$	”
93	282	$6^2 + 93^2 + 282^2 = 297^2$	$2^2 + 31^2 + 94^2 = 99^2$	”

etc. //

Theorem 16.3. For $a = 9$, there hold the identities for integral values of n, b.

n	b	Identity	Equivalently	Ref.
18	6	$9^2 + 18^2 + 6^2 = 21^2$	$3^2 + 6^2 + 2^2 = 7^2$	Th. 2.2
42	54	$9^2 + 42^2 + 54^2 = 69^2$	$3^2 + 14^2 + 18^2 = 23^2$	Th. 6.3
48	72	$9^2 + 48^2 + 72^2 = 87^2$	$3^2 + 16^2 + 24^2 = 29^2$	”
72	168	$9^2 + 72^2 + 168^2 = 183^2$	$3^2 + 24^2 + 56^2 = 61^2$	”
78	198	$9^2 + 78^2 + 198^2 = 213^2$	$3^2 + 26^2 + 66^2 = 71^2$	”
102	342	$9^2 + 102^2 + 342^2 = 357^2$	$3^2 + 34^2 + 114^2 = 119^2$	”

etc. //

Theorem 16.4. For $a = 12$, there hold the identities for integral values of n, b.

n	b	Identity	Equivalently	Ref.
21	12	$12^2 + 21^2 + 12^2 = 27^2$	$4^2 + 7^2 + 4^2 = 9^2$	Th. 3.2
39	48	$12^2 + 39^2 + 48^2 = 63^2$	$4^2 + 13^2 + 16^2 = 21^2$	Th. 6.4
51	84	$12^2 + 51^2 + 84^2 = 99^2$	$4^2 + 17^2 + 28^2 = 33^2$	”
69	156	$12^2 + 69^2 + 156^2 = 171^2$	$4^2 + 23^2 + 52^2 = 57^2$	”
81	216	$12^2 + 81^2 + 216^2 = 231^2$	$4^2 + 27^2 + 72^2 = 77^2$	”
99	324	$12^2 + 99^2 + 324^2 = 339^2$	$4^2 + 33^2 + 108^2 = 113^2$	”

etc. //

Theorem 16.5. For $a = 15$, there hold the identities for integral values of n, b.

n	b	Identity	Equivalently	Ref.
30	30	$15^2 + 30^2 + 30^2 = 45^2$	$1^2 + 2^2 + 2^2 = 3^2$	Th. 2.1
60	120	$15^2 + 60^2 + 120^2 = 135^2$	$1^2 + 4^2 + 8^2 = 9^2$	"
90	270	$15^2 + 90^2 + 270^2 = 285^2$	$1^2 + 6^2 + 18^2 = 19^2$	"
120	480	$15^2 + 120^2 + 480^2 = 495^2$	$1^2 + 8^2 + 32^2 = 33^2$	"
150	750	$15^2 + 150^2 + 750^2 = 765^2$	$1^2 + 10^2 + 50^2 = 51^2$	"

etc. //

Theorem 16.6. For $a = 18$, there hold the identities for integral values of n, b.

n	b	Identity	Equivalently	Ref.
9	6	$18^2 + 9^2 + 6^2 = 21^2$	$6^2 + 3^2 + 2^2 = 7^2$	Th. 2.2
21	18	$18^2 + 21^2 + 18^2 = 33^2$	$6^2 + 7^2 + 6^2 = 11^2$	Th. 5.3
39	54	$18^2 + 39^2 + 54^2 = 69^2$	$6^2 + 13^2 + 18^2 = 23^2$	Th. 6.6
51	90	$18^2 + 51^2 + 90^2 = 105^2$	$6^2 + 17^2 + 30^2 = 35^2$	"
69	162	$18^2 + 69^2 + 162^2 = 177^2$	$6^2 + 23^2 + 54^2 = 59^2$	"
81	222	$18^2 + 81^2 + 222^2 = 237^2$	$6^2 + 27^2 + 74^2 = 79^2$	"

etc. //

Theorem 16.7. For $a = 21$, there hold the identities for integral values of n, b.

n	b	Identity	Equivalently	Ref.
12	12	$21^2 + 12^2 + 12^2 = 27^2$	$7^2 + 4^2 + 4^2 = 9^2$	Th. 3.2
18	18	$21^2 + 18^2 + 18^2 = 33^2$	$7^2 + 6^2 + 6^2 = 11^2$	Th. 5.3
42	66	$21^2 + 42^2 + 66^2 = 81^2$	$7^2 + 14^2 + 22^2 = 27^2$	Th. 6.7
48	84	$21^2 + 48^2 + 84^2 = 99^2$	$7^2 + 16^2 + 28^2 = 33^2$	”
72	180	$21^2 + 72^2 + 180^2 = 195^2$	$7^2 + 24^2 + 60^2 = 65^2$	”
78	210	$21^2 + 78^2 + 210^2 = 225^2$	$7^2 + 26^2 + 70^2 = 75^2$	”

etc. //

§ 17. Identities of the type $a^2 + n^2 + b^2 = (b + 16)^2$

Above type of identities require:

$$b = (a^2 + n^2 - 256) / 32 = (a^2 + n^2) / 32 - 8, \qquad (17.1)$$

where a and n assume some suitable integral values making b integer. For positive integral values of a above relation yields the following values of b:

a	b	Remark
1	$(n^2 + 1) / 32 - 8$	b is not integer for any integer n.
2	$(n^2 + 4) / 32 - 8$	”
3	$(n^2 + 9) / 32 \quad 8$	”
4	$(n^2 - 16) / 32 - 7$	b is +ve integer for n = 20, 28, 36, 44, 52, 60, etc.
5	$(n^2 - 231)/32 = (n^2 - 7)/32 - 7$	b is not integer for any integer n.

6	$(n^2 - 220)/32 = (n^2 + 4)/32 - 7$	"
7	$(n^2 - 207)/32 = (n^2 - 15)/32 - 6$	"
8	$(n^2 - 192)/32 = n^2 / 32 - 6$	b is integer for n is any integral multiple of 8.
9	$(n^2 - 175)/32 = (n^2 - 15)/32 - 5$	b is not integer for any integer n.
10	$(n^2 - 156)/32 = (n^2 + 4)/32 - 5$	"
11	$(n^2 - 135)/32 = (n^2 - 7)/32 - 4$	"
12	$(n^2 - 112)/32 = (n^2 - 16)/32 - 3$	As for $a = 4$.
13	$(n^2 - 87)/32 = (n^2 + 9)/32 - 3$	b is not integer for any integer n.
14	$(n^2 - 60)/32 = (n^2 + 4)/32 - 2$	"
15	$(n^2 - 31)/32 = (n^2 + 1)/32 - 1$	"
16	$n^2 / 32$	As for $a = 8$.
17	$(n^2 + 33)/32 = (n^2 + 1)/32 + 1$	As for $a = 15$.
18	$(n^2 + 68)/32 = (n^2 + 4)/32 + 2$	As for $a = 14$.
19	$(n^2 + 105)/32 = (n^2 + 9)/32 + 3$	As for $a = 13$.
20	$(n^2 + 144)/32 = (n^2 - 16)/32 + 5$	As for $a = 4$.

etc. Conclusively, b takes integral values whenever a is an integral multiple of 4. Hence, there hold the following theorems:

Theorem 17.1. For $a = 4 \Rightarrow b = (n^2 - 16)/32 - 7$, there hold the identities for integral values of n, b.

n	b	Identity	Ref.
20	5	$4^2 + 20^2 + 5^2 = 21^2$	Th. 2.4

28	17	$4^2 + 28^2 + 17^2 = 33^2$	Th. 6.4
36	33	$4^2 + 36^2 + 33^2 = 49^2$	Th. 14.4
44	53	$4^2 + 44^2 + 53^2 = 69^2$	
52	77	$4^2 + 52^2 + 77^2 = 93^2$	
60	105	$4^2 + 60^2 + 105^2 = 121^2$	

etc. //

Theorem 17.2. For $a = 8 \Rightarrow b = n^2/32 - 6$, there hold the identities for positive integral values of n, b.

n	b	Identity	Equivalently	Ref.
16	2	$8^2 + 16^2 + 2^2 = 18^2$	$4^2 + 8^2 + 1^2 = 9^2$	Th. 2.1
24	12	$8^2 + 24^2 + 12^2 = 28^2$	$2^2 + 6^2 + 3^2 = 7^2$	Th. 2.2
32	26	$8^2 + 32^2 + 26^2 = 42^2$	$4^2 + 16^2 + 13^2 = 21^2$	Th. 6.4
40	44	$8^2 + 40^2 + 44^2 = 60^2$	$2^2 + 10^2 + 11^2 = 15^2$	Th. 5.1
48	66	$8^2 + 48^2 + 66^2 = 82^2$	$4^2 + 24^2 + 33^2 = 41^2$	Th. 9.1
56	92	$8^2 + 56^2 + 92^2 = 108^2$	$2^2 + 14^2 + 23^2 = 27^2$	Th. 5.1

etc. //

Theorem 17.3. For $a = 12 \Rightarrow b = (n^2 - 16)/32 - 3$, there hold the identities for integral values of n, b.

n	b	Identity	Ref.
12	1	$12^2 + 12^2 + 1^2 = 17^2$	Th. 6.1
20	9	$12^2 + 20^2 + 9^2 = 25^2$	Th. 6.9

28	21	$12^2 + 28^2 + 21^2 = 37^2$	Th. 10.4
36	37	$12^2 + 36^2 + 37^2 = 53^2$	
44	57	$12^2 + 44^2 + 57^2 = 73^2$	
52	81	$12^2 + 52^2 + 81^2 = 97^2$	

etc. //

Theorem 17.4. For $a = 16 \Rightarrow b = n^2 / 32$, there hold the identities for integral values of n, b.

n	b	Identity	Equivalently	Ref.
8	2	$16^2 + 8^2 + 2^2 = 18^2$	$8^2 + 4^2 + 1^2 = 9^2$	Th. 2.1
16	8	$16^2 + 16^2 + 8^2 = 24^2$	$2^2 + 2^2 + 1^2 = 3^2$	”
24	18	$16^2 + 24^2 + 18^2 = 34^2$	$8^2 + 12^2 + 9^2 = 17^2$	Th. 6.8
32	32	$16^2 + 32^2 + 32^2 = 48^2$	$1^2 + 2^2 + 2^2 = 3^2$	Th. 2.1
40	50	$16^2 + 40^2 + 50^2 = 66^2$	$8^2 + 20^2 + 25^2 = 33^2$	Th. 9.2
48	72	$16^2 + 48^2 + 72^2 = 88^2$	$2^2 + 6^2 + 9^2 = 11^2$	Th. 3.1

etc. //

Theorem 17.5. For $a = 20 \Rightarrow b = (n^2 - 16)/32 + 5$, there hold the identities for integral values of n, b.

n	b	Identity	Ref.
4	5	$20^2 + 4^2 + 5^2 = 21^2$	Theo. 2.4
12	9	$20^2 + 12^2 + 9^2 = 25^2$	Th. 6.9
20	17	$20^2 + 20^2 + 17^2 = 33^2$	§ 14

28	29	$20^2 + 28^2 + 29^2 = 45^2$	
36	45	$20^2 + 36^2 + 45^2 = 61^2$	
44	65	$20^2 + 44^2 + 65^2 = 81^2$	

etc. //

§ 18. Identities of the type $a^2 + n^2 + b^2 = (b + 17)^2$

Above type of identities require:

$$b = (a^2 + n^2 - 289)/34 = (a^2 + n^2 - 17)/34 - 8, \qquad (18.1)$$

where a and n assume some suitable integral values making b integer. For positive integral values of a above relation yields the following values of b:

a	b	Remark
1	$(n^2 - 16)/34 - 8$	b is +ve integer for $n = 30$, 38, 64, 72, 98, 106, etc.
2	$(n^2 - 13)/34 - 8$	b is +ve integer for $n = 25$, 43, 59, 77, 93, 111, etc.
3	$(n^2 - 8)/34 - 8$	b is +ve integer for $n = 22$, 46, 56, 80, 90, 114, etc.
4	$(n^2 - 1)/34 - 8$	b is +ve integer for $n = 33$, 35, 67, 69, 101, 103 etc.
5	$(n^2 + 8)/34 - 8$	b is +ve integer for $n = 20$, 48, 54, 82, 88, 116, etc.
6	$(n^2 - 253)/34 = (n^2 - 15)/34 - 7$	b is +ve integer for $n = 27$, 41, 61, 75, 95, 109, etc.
7	$(n^2 - 240)/34 = (n^2 - 2)/34 - 7$	b is +ve integer for $n = 28$, 40, 62, 74, 96, 108, etc.
8	$(n^2 - 225)/34 = (n^2 + 13)/34 - 7$	b is +ve integer for $n = 19$, 49, 53, 83, 87, 117, etc.
9	$(n^2 - 208)/34 = (n^2 - 4)/34 - 6$	b is integer when $n = 32$, 36, 66, 70, 100,104, etc.
10	$(n^2 - 189)/34 = (n^2 + 15)/34 - 6$	b is +ve integer for $n = 23$, 45, 57, 79, 91, 113, etc.

11	$(n^2 - 168)/34 = (n^2 + 2)/34 - 5$	b is +ve integer for n = 24, 44, 58, 78, 92, 112, etc.
12	$(n^2 - 145)/34 = (n^2 - 9)/34 - 4$	b is +ve integer for n = 31, 37, 65, 71, 99, 105, etc.
13	$(n^2 - 120)/34 = (n^2 + 16)/34 - 4$	b is +ve integer for n = 16, 18, 50, 52, 84, 86, etc.
14	$(n^2 - 93)/34 = (n^2 + 9)/34 - 3$	b is +ve integer for n = 29, 39, 63, 73, 97, 107, etc.
15	$(n^2 - 64)/34 = (n^2 + 4)/34 - 2$	b is +ve integer for n = 26, 42, 60, 76, 94, 110, etc.
16	$(n^2 - 33)/34 = (n^2 + 1)/34 - 1$	b is +ve integer for n = 13, 21, 47, 55, 81, 89, etc.
17	$n^2 / 34$	b is integer when n is an integral multiple of 34

etc. Thus, we have the:

Theorem 18.1. For $a = 1$, there hold the identities for integral values of n, b.

n	b	Identity	Ref.
30	18	$1^2 + 30^2 + 18^2 = 35^2$	Th. 6.1
38	34	$1^2 + 38^2 + 34^2 = 51^2$	Th. 14.1
64	112	$1^2 + 64^2 + 112^2 = 129^2$	
72	144	$1^2 + 72^2 + 144^2 = 161^2$	
98	174	$1^2 + 98^2 + 174^2 = 191^2$	
106	322	$1^2 + 106^2 + 322^2 = 339^2$	

etc. //

Theorem 18.2. For $a = 2$, there hold the identities for integral values of n, b.

n	b	Identity	Ref.
25	10	$2^2 + 25^2 + 10^2 = 27^2$	Th.3.1

43	46	$2^2 + 43^2 + 46^2 = 63^2$	
59	94	$2^2 + 59^2 + 94^2 = 111^2$	
77	166	$2^2 + 77^2 + 166^2 = 183^2$	
93	246	$2^2 + 93^2 + 246^2 = 263^2$	
111	354	$2^2 + 111^2 + 354^2 = 371^2$	

etc. //

Theorem 18.3. For $a = 3$, there hold the identities for integral values of n, b.

n	b	Identity	Ref.
22	6	$3^2 + 22^2 + 6^2 = 23^2$	Th. 2.3
46	54	$3^2 + 46^2 + 54^2 = 71^2$	
56	84	$3^2 + 56^2 + 84^2 = 101^2$	
80	180	$3^2 + 80^2 + 180^2 = 197^2$	
90	230	$3^2 + 90^2 + 230^2 = 247^2$	
114	374	$3^2 + 114^2 + 374^2 = 391^2$	

etc. //

Theorem 18.4. For $a = 4$, there hold the identities for integral values of n, b.

n	b	Identity	Ref.
33	24	$4^2 + 33^2 + 24^2 = 41^2$	Th. 9.1
35	28	$4^2 + 35^2 + 28^2 = 45^2$	Th. 11.2
67	124	$4^2 + 67^2 + 124^2 = 141^2$	

69	132	$4^2 + 69^2 + 132^2 = 149^2$	
101	292	$4^2 + 101^2 + 292^2 = 309^2$	
103	304	$4^2 + 103^2 + 304^2 = 321^2$	

etc. //

Theorem 18.5. For $a = 5$, there hold the identities for integral values of n, b.

n	b	Identity	Ref.
20	4	$5^2 + 20^2 + 4^2 = 21^2$	Th. 2.4
48	60	$5^2 + 48^2 + 60^2 = 77^2$	
54	78	$5^2 + 54^2 + 78^2 = 95^2$	
82	190	$5^2 + 82^2 + 190^2 = 207^2$	
88	220	$5^2 + 88^2 + 220^2 = 237^2$	
116	388	$5^2 + 116^2 + 388^2 = 405^2$	

etc. //

Theorem 18.6. For $a = 6$, there hold the identities for integral values of n, b.

n	b	Identity	Ref.
27	14	$6^2 + 27^2 + 14^2 = 31^2$	Th. 5.3
41	42	$6^2 + 41^2 + 42^2 = 59^2$	
61	102	$6^2 + 61^2 + 102^2 = 119^2$	
75	158	$6^2 + 75^2 + 158^2 = 175^2$	
95	258	$6^2 + 95^2 + 258^2 = 275^2$	

109	342	$6^2 + 109^2 + 342^2 = 359^2$	

etc. //

Theorem 18.7. For $a = 7$, there hold the identities for integral values of n, b.

n	b	Identity	Ref.
28	16	$7^2 + 28^2 + 16^2 = 33^2$	Th. 6.7
40	40	$7^2 + 40^2 + 40^2 = 57^2$	
62	106	$7^2 + 62^2 + 106^2 = 123^2$	
74	154	$7^2 + 74^2 + 154^2 = 171^2$	
96	264	$7^2 + 96^2 + 264^2 = 281^2$	
108	336	$7^2 + 108^2 + 336^2 = 353^2$	

etc. //

Theorem 18.8. For $a = 8$, there hold the identities for integral values of n, b.

n	b	Identity	Ref.
19	4	$8^2 + 19^2 + 4^2 = 21^2$	Th. 3.2
49	64	$8^2 + 49^2 + 64^2 = 81^2$	
53	76	$8^2 + 53^2 + 76^2 = 93^2$	
83	196	$8^2 + 83^2 + 196^2 = 213^2$	
87	216	$8^2 + 87^2 + 216^2 = 233^2$	
117	396	$8^2 + 117^2 + 396^2 = 413^2$	

etc. //

Theorem 18.9. For $a = 9$, there hold the identities for integral values of n, b.

n	b	Identity	Ref.
32	24	$9^2 + 32^2 + 24^2 = 41^2$	Th. 10.3
36	32	$9^2 + 36^2 + 32^2 = 49^2$	Th. 14.9
66	122	$9^2 + 66^2 + 122^2 = 139^2$	
70	138	$9^2 + 70^2 + 138^2 = 155^2$	
100	288	$9^2 + 100^2 + 288^2 = 305^2$	
104	312	$9^2 + 104^2 + 312^2 = 329^2$	

etc. //

Theorem 18.10. For $a = 10$, there hold the identities for integral values of n, b.

n	b	Identity	Ref.
23	10	$10^2 + 23^2 + 10^2 = 27^2$	Th. 5.5
45	54	$10^2 + 45^2 + 54^2 = 71^2$	
57	90	$10^2 + 57^2 + 90^2 = 107^2$	
79	178	$10^2 + 79^2 + 178^2 = 195^2$	
91	238	$10^2 + 91^2 + 238^2 = 255^2$	
113	370	$10^2 + 113^2 + 370^2 = 387^2$	

etc. //

Theorem 18.11. For $a = 11$, there hold the identities for integral values of n, b.

n	b	Identity	Ref.
24	12	$11^2 + 24^2 + 12^2 = 29^2$	§ 6
44	52	$11^2 + 44^2 + 52^2 = 69^2$	
58	94	$11^2 + 58^2 + 94^2 = 111^2$	
78	174	$11^2 + 78^2 + 174^2 = 191^2$	
92	244	$11^2 + 92^2 + 244^2 = 261^2$	
112	364	$11^2 + 112^2 + 364^2 = 381^2$	

etc. //

Theorem 18.12. For $a = 12$, there hold the identities for integral values of n, b.

n	b	Identity	Ref.
31	24	$12^2 + 31^2 + 24^2 = 41^2$	§ 11
37	36	$12^2 + 37^2 + 36^2 = 53^2$	Th. 17.3
65	120	$12^2 + 65^2 + 120^2 = 137^2$	
71	144	$12^2 + 71^2 + 144^2 = 161^2$	
99	284	$12^2 + 99^2 + 284^2 = 301^2$	
105	320	$12^2 + 105^2 + 320^2 = 337^2$	

etc. //

Theorem 18.13. For $a = 13$, there hold the identities for integral values of n, b.

n	b	Identity	Ref.
16	4	$13^2 + 16^2 + 4^2 = 21^2$	Th. 6.4
18	6	$13^2 + 18^2 + 6^2 = 23^2$	Th. 6.6
50	70	$13^2 + 50^2 + 70^2 = 87^2$	
52	76	$13^2 + 52^2 + 76^2 = 93^2$	
84	204	$13^2 + 84^2 + 204^2 = 221^2$	
86	214	$13^2 + 86^2 + 214^2 = 231^2$	

etc. //

Theorem 18.14. For $a = 14$, there hold the identities for integral values of n, b.

n	b	Identity	Ref.
29	22	$14^2 + 29^2 + 22^2 = 39^2$	§ 11
39	42	$14^2 + 39^2 + 42^2 = 59^2$	
63	114	$14^2 + 63^2 + 114^2 = 131^2$	
73	154	$14^2 + 73^2 + 154^2 = 171^2$	
97	274	$14^2 + 97^2 + 274^2 = 291^2$	
107	334	$14^2 + 107^2 + 334^2 = 351^2$	

etc. //

Theorem 18.15. For $a = 15$, there hold the identities for integral values of n, b.

n	b	Identity	Ref.
26	18	$15^2 + 26^2 + 18^2 = 35^2$	§ 10
42	50	$15^2 + 42^2 + 50^2 = 67^2$	
60	104	$15^2 + 60^2 + 104^2 = 121^2$	
76	168	$15^2 + 76^2 + 168^2 = 185^2$	
94	258	$15^2 + 94^2 + 258^2 = 275^2$	
110	354	$15^2 + 110^2 + 354^2 = 371^2$	

etc. //

Theorem 18.16. For $a = 16$, there hold the identities for integral values of n, b.

n	b	Identity	Ref.
13	4	$16^2 + 13^2 + 4^2 = 21^2$	Th. 6.4
21	12	$16^2 + 21^2 + 12^2 = 29^2$	Th. 9.3
47	64	$16^2 + 47^2 + 64^2 = 81^2$	
55	88	$16^2 + 55^2 + 88^2 = 105^2$	
81	192	$16^2 + 81^2 + 192^2 = 209^2$	
89	232	$16^2 + 89^2 + 232^2 = 249^2$	

etc. //

Theorem 18.17. For $a = 17$, there hold the identities for integral values of n, b.

n	b	Identity	Equivalently	Ref.
34	34	$17^2 + 34^2 + 34^2 = 51^2$	$1^2 + 2^2 + 2^2 = 3^2$	Th. 2.1
68	136	$17^2 + 68^2 + 136^2 = 153^2$	$1^2 + 4^2 + 8^2 = 9^2$	”
102	306	$17^2 + 102^2 + 306^2 = 323^2$	$1^2 + 6^2 + 18^2 = 19^2$	”
136	544	$17^2 + 136^2 + 544^2 = 561^2$	$1^2 + 8^2 + 32^2 = 33^2$	”
170	850	$17^2 + 170^2 + 850^2 = 867^2$	$1^2 + 10^2 + 50^2 = 51^2$	”
204	1224	$17^2 + 204^2 + 1224^2 = 1241^2$	$1^2 + 12^2 + 72^2 = 73^2$	”

etc. //

§ 19. Identities of complex nature

Theorem 19.1. Interestingly, there holds the following identity:

$$a^2 + (a + 1)^2 + \{a\,(a + 1)\}^2 = (a^2 + a + 1)^2, \qquad (19.1)$$

where a is any real number.

Proof. Left hand expression expands as

$$a^2 + (a^2 + 2a + 1) + (a^4 + 2a^3 + a^2) = a^4 + 2a^3 + 3a^2 + 2a + 1,$$

and so is the right hand expression for any real value of a. However, confining to only positive integral values of a, one may derive the following identities:

a	Identity	Ref.
1	$1^2 + 2^2 + 2^2 = 3^2$	Th. 2.1
2	$2^2 + 3^2 + 6^2 = 7^2$	Th. 2.2

3	$3^2 + 4^2 + 12^2 = 13^2$	Th. 2.3
4	$4^2 + 5^2 + 20^2 = 21^2$	Th. 2.4
5	$5^2 + 6^2 + 30^2 = 31^2$	Th. 2.5
6	$6^2 + 7^2 + 42^2 = 43^2$	Th. 2.6

etc. //

Theorem 19.2. Another set of identity satisfied by some integers could be of the type:

$$a^2 + (2n)^2 + b^2 = (a + b)^2, \qquad (19.2)$$

when b must be equal to $2n^2 / a$.

Proof. Expanding the right hand expression and cancelling the common terms on either side one easily derives the value of b as above. //

Giving different positive integral values to a and n, we thus derive the following identities:

Theorem 19.3. Choosing $a = 1 \Rightarrow b = 2n^2$, there hold the identities

n	b	Identity	Ref.
1	2	$1^2 + 2^2 + 2^2 = 3^2$	Th. 2.1
2	8	$1^2 + 4^2 + 8^2 = 9^2$	”
3	18	$1^2 + 6^2 + 18^2 = 19^2$	”
4	32	$1^2 + 8^2 + 32^2 = 33^2$	”
5	50	$1^2 + 10^2 + 50^2 = 51^2$	”
6	72	$1^2 + 12^2 + 72^2 - 73^2$	”

etc. //

Theorem 19.4. Choosing $a = 2 \Rightarrow b = n^2$, there hold the identities

n	b	Identity	Equivalently	Ref.
1	1	$2^2 + 2^2 + 1^2 = 3^2$		Th. 2.1
2	4	$2^2 + 4^2 + 4^2 = 6^2$	$1^2 + 2^2 + 2^2 = 3^2$	,,
3	9	$2^2 + 6^2 + 9^2 = 11^2$		Th. 3.1
4	16	$2^2 + 8^2 + 16^2 = 18^2$	$1^2 + 4^2 + 8^2 = 9^2$	Th. 2.1
5	25	$2^2 + 10^2 + 25^2 = 27^2$		Th. 3.1
6	36	$2^2 + 12^2 + 36^2 = 38^2$	$1^2 + 6^2 + 18^2 = 19^2$	Th. 2.1

etc. //

Theorem 19.5. Choosing $a = 3 \Rightarrow b = 2n^2/3$, there hold the identities

n	b	Identity	Equivalently	Ref.
3	6	$3^2 + 6^2 + 6^2 = 9^2$	$1^2 + 2^2 + 2^2 = 3^2$	Th. 2.1
6	24	$3^2 + 12^2 + 24^2 = 27^2$	$1^2 + 4^2 + 8^2 = 9^2$	,,
9	54	$3^2 + 18^2 + 54^2 = 57^2$	$1^2 + 6^2 + 18^2 = 19^2$	,,
12	96	$3^2 + 24^2 + 96^2 = 99^2$	$1^2 + 8^2 + 32^2 = 33^2$	,,
15	150	$3^2 + 30^2 + 150^2 = 153^2$	$1^2 + 10^2 + 50^2 = 51^2$	,,
18	216	$3^2 + 36^2 + 216^2 = 219^2$	$1^2 + 12^2 + 72^2 = 73^2$	,,

etc. //

Unlike to Eq. (19.2) next set of identities may be considered as

$$a^2 + (2n + 1)^2 + b^2 = (1 + b)^2, \tag{19.3}$$

when b must be equal to $a^2/2 + 2n\,(n + 1)$.

Proof. Expanding the right hand expression and cancelling the common terms on either side we easily derive the value of b as above. //

In order to make b whole number we choose even integral values of a and give different positive integral values to n. We thus derive the following identities:

Theorem 19.6. Choosing $a = 2 \Rightarrow b = 2 + 2n\,(n + 1)$, there hold the identities

n	$n\,(n+1)$	b	Identity	Ref.
1	2	6	$2^2 + 3^2 + 6^2 = 7^2$	Th. 2.2
2	6	14	$2^2 + 5^2 + 14^2 = 15^2$	,,
3	12	26	$2^2 + 7^2 + 26^2 = 27^2$	,,
4	20	42	$2^2 + 9^2 + 42^2 = 43^2$	,,
5	30	62	$2^2 + 11^2 + 62^2 = 63^2$	,,
6	42	86	$2^2 + 13^2 + 86^2 = 87^2$	,,

etc. //

Theorem 19.7. Choosing $a = 4 \Rightarrow b = 8 + 2n\,(n + 1)$, there hold the identities

n	$n\,(n+1)$	b	Identity	Ref.
1	2	12	$4^2 + 3^2 + 12^2 = 13^2$	Th. 2.3
2	6	20	$4^2 + 5^2 + 20^2 = 21^2$	Th. 2.4
3	12	32	$4^2 + 7^2 + 32^2 = 33^2$	,,
4	20	48	$4^2 + 9^2 + 48^2 = 49^2$	,,
5	30	68	$4^2 + 11^2 + 68^2 = 69^2$	,,

6	42	92	$4^2 + 13^2 + 92^2 = 93^2$	"

etc. //

Exploring possibilities so that the identity

$$a^2 + (n + 2)^2 + b^2 = (2 + b)^2$$

holds for some integral values of a and n, we must have

$$b = (a^2 + n^2) / 4 + n.$$

Proof. Expanding the terms on both sides and cancelling the common terms one easily derives the value of b. //

Theorem 19.8. Choosing $a = 2 \Rightarrow b = n^2/4 + n + 1$, there hold the identities:

n	b	Identity	Equivalently	Ref.
2	4	$2^2 + 4^2 + 4^2 = 6^2$	$1^2 + 2^2 + 2^2 = 3^2$	Th. 2.1
4	9	$2^2 + 6^2 + 9^2 = 11^2$		Th. 3.1
6	16	$2^2 + 8^2 + 16^2 = 18^2$	$1^2 + 4^2 + 8^2 = 9^2$	Th. 2.1
8	25	$2^2 + 10^2 + 25^2 = 27^2$		Th. 3.1
10	36	$2^2 + 12^2 + 36^2 = 38^2$	$1^2 + 6^2 + 18^2 = 19^2$	Th. 2.1

etc. It may be noted that these identities have also been found in Theo. 19.4. //

Theorem 19.9. Choosing $a = 4 \Rightarrow b = n^2/4 + n + 4$. There hold the identities:

n	b	Identity	Equivalently	Ref.
2	7	$4^2 + 4^2 + 7^2 = 9^2$		Th. 3.2

4	12	$4^2 + 6^2 + 12^2 = 14^2$	$2^2 + 3^2 + 6^2 = 7^2$	Th. 2.2
6	19	$4^2 + 8^2 + 19^2 = 21^2$		Th. 3.2
8	28	$4^2 + 10^2 + 28^2 = 30^2$	$2^2 + 5^2 + 14^2 = 15^2$	Th. 2.2
10	39	$4^2 + 12^2 + 39^2 = 41^2$		Th. 3.2

etc. //

Theorem 19.10. Choosing $a = 6 \Rightarrow b = n^2/4 + n + 9$, there hold the identities:

n	b	Identity	Equivalently	Ref.
2	12	$6^2 + 4^2 + 12^2 = 14^2$	$3^2 + 2^2 + 6^2 = 7^2$	Th. 2.2
4	17	$6^2 + 6^2 + 17^2 = 19^2$		Th. 3.3
6	24	$6^2 + 8^2 + 24^2 = 26^2$	$3^2 + 4^2 + 12^2 = 13^2$	Th. 2.3
8	33	$6^2 + 10^2 + 33^2 = 35^2$		Th. 3.3
10	44	$6^2 + 12^2 + 44^2 = 46^2$	$3^2 + 6^2 + 22^2 = 23^2$	Th. 2.3

etc. //

Theorem 19.11. Choosing $a = 8 \Rightarrow b = n^2/4 + n + 16$, there hold the identities:

n	b	Identity	Equivalently	Ref.
2	19	$8^2 + 4^2 + 19^2 = 21^2$		Th. 3.2
4	24	$8^2 + 6^2 + 24^2 = 26^2$	$4^2 + 3^2 + 12^2 = 13^2$	Th. 2.3
6	31	$8^2 + 8^2 + 31^2 = 33^2$		Th. 3.4
8	40	$8^2 + 10^2 + 40^2 = 42^2$	$4^2 + 5^2 + 20^2 = 21^2$	Th. 2.4
10	51	$8^2 + 12^2 + 51^2 = 53^2$		Th. 3.4

etc. //

Note 19.1. Many results of the paper were worked out during author's stay at Divine Word Univ., Madang (PNG).

CHAPTER 2

GENERALIZATIONS OF PYTHAGORAS THEOREM TO QUADRILATERALS - II

The first eighteen Sections in previous chapter dealt with the direct sum of squares of three positive integers making the square of the fourth integer. Special choices for $d = c + k$, where k ranged over the set of natural numbers from 1 to 17 were discussed. Presently, we continue with the discussion for further integral powers of $k = 18$ onward.

§ 1. Identities of the type $a^2 + n^2 + b^2 = (b + 18)^2$

Above type of identities require:

$$b = (a^2 + n^2 - 18^2) / 36 = (a^2 + n^2) / 36 - 9. \qquad (1.1)$$

To get integral values of b the sum $a^2 + n^2$ must be divisible by 36, which is possible only when both a and n must be integral multiples of 6: say $a = 6\,l$ and $n = 6\,m$, where both l, m are integers making

$$(a^2 + n^2) / 36 = l^2 + m^2 \;\Rightarrow\; b = l^2 + m^2 - 9 \qquad (1.2)$$

Thus, the different integral values of l, m yield different integral values to b and, therefore, to $b + 18$. The following tables depict such choices and yield the corresponding identities of the desired form:

Theorem 1.1. For $l = 1$ so that $a = 6$ and $b = m^2 - 8$, there hold the identities for integral values of n, b.

m	$n = 6m$	b	Identity	Ref. Chapt.1
1	6	-7	$6^2 + 6^2 + (-7)^2 = 11^2$	Th. 5.3
2	12	-4	$6^2 + 12^2 + (-4)^2 = 14^2$	Th. 3.2
			i.e. $3^2 + 6^2 + (-2)^2 = 7^2$	Th. 2.2
3	18	1	$6^2 + 18^2 + 1^2 = 19^2$	Th. 2.1

4	24	8	$6^2 + 24^2 + 8^2 = 26^2$	Th. 3.3
			i.e. $3^2 + 12^2 + 4^2 = 13^2$	Th. 2.3
5	30	17	$6^2 + 30^2 + 17^2 = 35^2$	Th. 6.6
6	36	28	$6^2 + 36^2 + 28^2 = 46^2$	Th. 11.3
			i.e. $3^2 + 18^2 + 14^2 = 23^2$	Th. 6.3
7	42	41	$6^2 + 42^2 + 41^2 = 59^2$	Th. 18.6
8	48	56	$6^2 + 48^2 + 56^2 = 74^2$	
			i.e. $3^2 + 24^2 + 28^2 = 37^2$	Th. 10.1
9	54	73	$6^2 + 54^2 + 73^2 = 91^2$	
10	60	92	$6^2 + 60^2 + 92^2 = 110^2$	
			i.e. $3^2 + 30^2 + 46^2 = 55^2$	Th. 10. 1

etc. //

Theorem 1.2. For $l = 2$ so that $a = 12$ and $b = m^2 - 5$, there hold the identities for integral values of n, b.

m	$n = 6m$	b	Identity	Ref. Chapt.1
2	12	– 1	$12^2 + 12^2 + (-1)^2 = 17^2$	Th. 6.1
3	18	4	$12^2 + 18^2 + 4^2 = 22^2$	Th. 5.2
			i.e. $6^2 + 9^2 + 2^2 = 11^2$	Th. 3.1
4	24	11	$12^2 + 24^2 + 11^2 = 29^2$	
5	30	20	$12^2 + 30^2 + 20^2 = 38^2$	Th. 9.3
			i.e. $6^2 + 15^2 + 10^2 = 19^2$	Th. 5.3

6	36	31	$12^2 + 36^2 + 31^2 = 49^2$	Th. 14.12
7	42	44	$12^2 + 42^2 + 44^2 = 62^2$	
			i.e. $6^2 + 21^2 + 22^2 = 31^2$	Th. 10.2
8	48	59	$12^2 + 48^2 + 59^2 = 77^2$	
9	54	76	$12^2 + 54^2 + 76^2 = 94^2$	
			i.e. $6^2 + 27^2 + 38^2 = 47^2$	Th. 10.2
10	60	95	$12^2 + 60^2 + 95^2 = 113^2$	

etc. It may be noted that we have dropped the case $a = 12$, $n = 6$ leading to same value of $b = -4$ already considered in Theorem 1.1. //

Theorem 1.3. Taking $l = 3$ so that $a = 18$, and $b = m^2$ by Eq. (1.2). Hence, there hold the following identities for integral values of n, b.

m	$n = 6m$	b	Identity	Ref. Chap.1
3	18	9	$18^2 + 18^2 + 9^2 = 27^2$	Th. 10.3
			i.e. $2^2 + 2^2 + 1^2 = 3^2$	Th. 2.1
4	24	16	$18^2 + 24^2 + 16^2 = 34^2$	
			i.e. $9^2 + 12^2 + 8^2 = 17^2$	Th. 6.8
5	30	25	$18^2 + 30^2 + 25^2 = 43^2$	
6	36	36	$18^2 + 36^2 + 36^2 = 54^2$	
			i.e. $1^2 + 2^2 + 2^2 = 3^2$	Th. 2.1
7	42	49	$18^2 + 42^2 + 49^2 = 67^2$	
8	48	64	$18^2 + 48^2 + 64^2 = 82^2$	

			i.e. $9^2 + 24^2 + 32^2 = 41^2$	Th. 10.3
9	54	81	$18^2 + 54^2 + 81^2 = 99^2$	
			i.e. $2^2 + 6^2 + 9^2 = 11^2$	Th. 3.1
10	60	100	$18^2 + 60^2 + 100^2 = 118^2$	
			i.e. $9^2 + 30^2 + 50^2 = 59^2$	Th. 10. 3

etc. Here also, the pairs $(a, n) = (18, 6)$ and $(18, 12)$ have been dropped as they are already considered in Theorems 1.1 1nd 1.2 respectively. //

Theorem 1.4. Taking $l = 4$ so that $a = 24$, and $b = m^2 + 7$ by Eq. (1.2). Hence, there hold the following identities for integral values of n, b.

m	$n = 6m$	b	Identity	Ref. Chap.1
4	24	23	$24^2 + 24^2 + 23^2 = 41^2$	
5	30	32	$24^2 + 30^2 + 32^2 = 50^2$	
			i.e. $12^2 + 15^2 + 16^2 = 25^2$	Th. 10.4
6	36	43	$24^2 + 36^2 + 43^2 = 61^2$	
7	42	56	$24^2 + 42^2 + 56^2 = 74^2$	
			i.e. $12^2 + 21^2 + 28^2 = 37^2$	Th. 10.4
8	48	71	$24^2 + 48^2 + 71^2 = 89^2$	
9	54	88	$24^2 + 54^2 + 88^2 = 106^2$	
			i.e. $12^2 + 27^2 + 44^2 = 53^2$	Th. 10.4
10	60	107	$24^2 + 60^2 + 107^2 = 125^2$	

etc. Here also, the values of a and n in the pairs $(a, n) = (24, 6), (24, 12)$ and $(24, 18)$ have been dropped as they are already included in previous Theorems. //

Theorem 1.5. Taking $l = 5$ so that $a = 30$, and $b = m^2 + 16$ by Eq. (1.2). Hence, there hold the following identities for integral values of n, b.

m	$n = 6m$	b	Identity	Ref. Chap.1
5	30	41	$30^2 + 30^2 + 41^2 = 59^2$	
6	36	52	$30^2 + 36^2 + 52^2 = 70^2$	
			i.e. $15^2 + 18^2 + 26^2 = 35^2$	§ 10
7	42	65	$30^2 + 42^2 + 65^2 = 83^2$	
8	48	80	$30^2 + 48^2 + 80^2 = 98^2$	
			i.e. $15^2 + 24^2 + 40^2 = 49^2$	§ 10
9	54	97	$30^2 + 54^2 + 97^2 = 115^2$	
10	60	116	$30^2 + 60^2 + 116^2 = 134^2$	
			i.e. $15^2 + 30^2 + 58^2 = 67^2$	§ 10

§ 2. Identities of the type $a^2 + n^2 + b^2 = (b + 19)^2$

Above type of identities require:

$$b = (a^2 + n^2 - 19^2) / 38 = \{ (a - 19).(a + 19) + n^2 \} / 38.$$

$$= (a^2 + n^2 - 19) / 38 - 9 \qquad (2.1)$$

To get integral values of b the sum $a^2 + n^2 - 19$ must be divisible by 38, i.e. it should be an integral multiple of 38. The possibility for vanishing of the sum $a^2 + n^2 - 19$ is nil as no integral values of a make n integer. Same is the situation for $a^2 + n^2 = 38k + 19$ for some integer k. We explore the following situations arising out from Eq. (2.1):

Table 2.1.

a	b	Remark
19	$n^2 / 38$	b is integer when n is an even integral multiple of 19.
38	$(n^2 + 19 \times 57) / 38$ $= (n^2 + 19) / 38 + 28$	b is integer when n is an odd integral multiple of 19.
57	$(n^2 + 38 \times 76) / 38$ $= n^2 / 38 + 76$	As for $a = 19$.
76	$(n^2 + 57 \times 95) / 38$ $= (n^2 + 19) / 38 + 142$	As for $a = 38$.
95	$(n^2 + 76 \times 114) / 38$ $= n^2 / 38 + 228$	As for $a = 19$.
114	$(n^2 + 95 \times 133) / 38$ $= (n^2 + 19) / 38 + 332$	As for $a = 38$.
133	$(n^2 + 114 \times 152) / 38$ $= n^2 / 38 + 456$	As for $a = 19$.
152	$(n^2 + 133 \times 171) / 38$ $= (n^2 + 19) / 38 + 598$	As for $a = 38$.
171	$(n^2 + 152 \times 190) / 38$ $= n^2 / 38 + 760$	As for $a = 19$.
190	$(n^2 + 171 \times 209) / 38$ $= (n^2 + 19) / 38 + 940$	As for $a = 38$.

etc. Conclusively, b becomes integer when a takes *odd* integral multiples of 19 together with corresponding n as *even* integral multiples of 19 and vice-versa, i.e. when a assuming even but n taking odd integral multiples of 19. Thus, we have the following theorems. Symmetry of a and n in the sum $a^2 + n^2$ is also noticeable hence, the pairs (a, n) and (n, a) shall yield the same values of b.

Theorem 2.1. For $a = 19$, $b = n^2 / 38$, by Table 2.1. Hence, $n = 38$, 76, 114, 152, etc. yield the following identities.

n	b	Identity	Refer Chap.1
38	38	$19^2 + 38^2 + 38^2 = 57^2$	
		i.e. $1^2 + 2^2 + 2^2 = 3^2$	Th. 2.1
76	152	$19^2 + 76^2 + 152^2 = 171^2$	
		i.e. $1^2 + 4^2 + 8^2 = 9^2$	Th. 2.1
114	342	$19^2 + 114^2 + 342^2 = 361^2$	
		i.e. $1^2 + 6^2 + 18^2 = 19^2$	Th. 2.1
152	608	$19^2 + 152^2 + 608^2 = 627^2$	
		i.e. $1^2 + 8^2 + 32^2 = 33^2$	Th. 2.1
190	950	$19^2 + 190^2 + 950^2 = 969^2$	
		i.e. $1^2 + 10^2 + 50^2 = 51^2$	Th. 2.1
228	1368	$19^2 + 228^2 + 1368^2 = 1387^2$	
		i.e. $1^2 + 12^2 + 72^2 = 73^2$	Th. 2.1
266	1862	$19^2 + 266^2 + 1862^2 = 1881^2$	
		i.e. $1^2 + 14^2 + 98^2 = 99^2$	Th. 2.1
304	2432	$19^2 + 304^2 + 2432^2 = 2451^2$	
		i.e. $1^2 + 16^2 + 128^2 = 129^2$	Th. 2.1
342	3078	$19^2 + 342^2 + 3078^2 = 3097^2$	
		i.e. $1^2 + 18^2 + 162^2 = 163^2$	Th. 2.1

380	3800	$19^2 + 380^2 + 3800^2 = 3819^2$	
		i.e. $1^2 + 20^2 + 200^2 = 201^2$	Th. 2.1

etc. //

Theorem 2.2. For $a = 38$, $b = (n^2 + 19)/38 + 28$ by Table 2.1. Hence, for $n = 57, 95, 133$ etc., we derive the following identities.

n	b	Identity	Ref. Chap.1
57	114	$38^2 + 57^2 + 114^2 = 133^2$	
		i.e. $2^2 + 3^2 + 6^2 = 7^2$	Th. 2.2
95	266	$38^2 + 95^2 + 266^2 = 285^2$	
		i.e. $2^2 + 5^2 + 14^2 = 15^2$	Th. 2.2
133	494	$38^2 + 133^2 + 494^2 = 513^2$	
		i.e. $2^2 + 7^2 + 26^2 = 27^2$	Th. 2.2
171	798	$38^2 + 171^2 + 798^2 = 817^2$	
		i.e. $2^2 + 9^2 + 42^2 = 43^2$	Th. 2.2
209	1178	$38^2 + 209^2 + 1178^2 = 1197^2$	
		i.e. $2^2 + 11^2 + 62^2 = 63^2$	Th. 2.2
247	1634	$38^2 + 247^2 + 1634^2 = 1653^2$	
		i.e. $2^2 + 13^2 + 86^2 = 87^2$	Th. 2.2
285	2166	$38^2 + 285^2 + 2166^2 = 2185^2$	
		i.e. $2^2 + 15^2 + 114^2 = 115^2$	Th. 2.2
323	2774	$38^2 + 323^2 + 2774^2 = 2793^2$	

		i.e. $2^2 + 17^2 + 146^2 = 147^2$	Th. 2.2
361	3458	$38^2 + 361^2 + 3458^2 = 3477^2$	
		i.e. $2^2 + 19^2 + 182^2 = 183^2$	Th. 2.2

etc. The choice of pair $(a, n) = (38, 19)$ is dropped as $(a, n) = (19, 38)$ yielding the same value of b is already considered in Theorem 2.1. //

Theorem 2.3. For $a = 57$, $b = n^2/38 + 76$, by Table 2.1. Hence, $n =$ 76, 114, 152, 190, etc. yield the following identities.

n	b	Identity	Ref. Chap.1
76	228	$57^2 + 76^2 + 228^2 = 247^2$	
		i.e. $3^2 + 4^2 + 12^2 = 13^2$	Th. 2.3
114	418	$57^2 + 114^2 + 418^2 = 437^2$	
		i.e. $3^2 + 6^2 + 22^2 = 23^2$	Th. 2.3
152	684	$57^2 + 152^2 + 684^2 = 703^2$	
		i.e. $3^2 + 8^2 + 36^2 = 37^2$	Th. 2.3
190	1026	$57^2 + 190^2 + 1026^2 = 1045^2$	
		i.e. $3^2 + 10^2 + 54^2 = 55^2$	Th. 2.3
228	1444	$57^2 + 228^2 + 1444^2 = 1463^2$	
		i.e. $3^2 + 12^2 + 76^2 = 77^2$	Th. 2.3
266	1938	$57^2 + 266^2 + 1938^2 = 1957^2$	
		i.e. $3^2 + 14^2 + 102^2 = 103^2$	Th. 2.3
304	2508	$57^2 + 304^2 + 2508^2 = 2527^2$	

		i.e. $3^2 + 16^2 + 132^2 = 133^2$	Th. 2.3
342	3154	$57^2 + 342^2 + 3154^2 = 3173^2$	
		i.e. $3^2 + 18^2 + 166^2 = 167^2$	Th. 2.3
380	3876	$57^2 + 380^2 + 3876^2 = 3895^2$	
		i.e. $3^2 + 20^2 + 204^2 = 205^2$	Th. 2.3

etc. The choice of pair $(a, n) = (57, 38)$ is dropped as $(a, n) = (38, 57)$ is already considered in Theorem 2.2. //

Theorem 2.4. For $a = 76$, $b = (n^2 + 19)/38 + 142$ by Table 2.1. Hence, $n = 95, 133, 171, 209, 247$, etc. yield the following identities.

n	b	Identity	Ref. Chap.1
95	380	$76^2 + 95^2 + 380^2 = 399^2$	
		i.e. $4^2 + 5^2 + 20^2 = 21^2$	Th. 2.2
133	608	$76^2 + 133^2 + 608^2 = 627^2$	
		i.e. $4^2 + 7^2 + 32^2 = 33^2$	Th. 2.2
171	912	$76^2 + 171^2 + 912^2 = 931^2$	
		i.e. $4^2 + 9^2 + 48^2 = 49^2$	Th. 2.2
209	1292	$76^2 + 209^2 + 1292^2 = 1311^2$	
		i.e. $4^2 + 11^2 + 68^2 = 69^2$	Th. 2.2
247	1748	$76^2 + 247^2 + 1748^2 = 1767^2$	
		i.e. $4^2 + 13^2 + 92^2 = 93^2$	Th. 2.2
285	2280	$76^2 + 285^2 + 2280^2 = 2299^2$	

		i.e. $4^2 + 15^2 + 120^2 = 121^2$	
323	2888	$76^2 + 323^2 + 2888^2 = 2907^2$	
		i.e. $4^2 + 17^2 + 152^2 = 153^2$	Th. 2.2
361	3572	$76^2 + 361^2 + 3572^2 = 3591^2$	
		i.e. $4^2 + 19^2 + 188^2 = 189^2$	Th. 2.2

etc. The choices $(a, n) = (76, 19)$ and $(76, 57)$ are dropped as the corresponding results are already considered in Theorems 2.1 and 2.3 respectively. //

§ 3. Identities of the type $a^2 + n^2 + b^2 = (b + 20)^2$

Above type of identities require:

$$b = (a^2 + n^2 - 20^2) / 40 = (a^2 + n^2) / 40 - 10. \quad (3.1)$$

To get integral values of b the sum $a^2 + n^2$ must be divisible by 40. The choice of paired values $(a, n) = (2k, 6k)$, where k is some integer make the possibility; for which Eq. (3.1) yields $b = k^2 - 10$. Thus we have the following Theorem.

Theorem 3.1. For $k = 1, 2, 3, \ldots$ there hold the identities

k	$a = 2k$	$n = 6k$	b	Identity	Ref. Chap.1
1	2	6	– 9	$2^2 + 6^2 + (-9)^2 = 11^2$	Th. 3.1
2	4	12	– 6	$4^2 + 12^2 + (-6)^2 = 14^2$	Th. 3.2
				i.e. $2^2 + 6^2 + (-3)^2 = 7^2$	Th. 2.2
3	6	18	– 1	$6^2 + 18^2 + (-1)^2 = 19^2$	Th. 2.1
4	8	24	6	$8^2 + 24^2 + 6^2 = 26^2$	Th. 3.3
				i.e. $4^2 + 12^2 + 3^2 = 13^2$	Th. 2.3

5	10	30	15	$10^2 + 30^2 + 15^2 = 35^2$	Th. 6.10
				i.e. $2^2 + 6^2 + 3^2 = 7^2$	Th. 2.2
6	12	36	26	$12^2 + 36^2 + 26^2 = 46^2$	§ 11
				i.e. $6^2 + 18^2 + 13^2 = 23^2$	Th. 6.6
7	14	42	39	$14^2 + 42^2 + 39^2 = 59^2$	Th. 18.14
8	16	48	54	$16^2 + 48^2 + 54^2 = 74^2$	
				i.e. $8^2 + 24^2 + 27^2 = 37^2$	Th. 11.4
9	18	54	71	$18^2 + 54^2 + 71^2 = 91^2$	
10	20	60	90	$20^2 + 60^2 + 90^2 = 110^2$	
				i.e. $2^2 + 6^2 + 9^2 = 11^2$	Th. 3.1

§ 4. Identities of the type $a^2 + n^2 + b^2 = (b + 21)^2$

Above type of identities require:

$$b = (a^2 + n^2 - 21^2) / 42 = \{ (a - 21).(a + 21) + n^2 \} / 42$$

$$= (a^2 + n^2 - 21) / 42 - 10 \qquad (4.1)$$

To get integral values of b the sum $a^2 + n^2 - 21$ must be divisible by 42. The possibility for vanishing of the sum $a^2 + n^2 - 21$ is nil as no integral values of a make n integer. Same is the situation for $a^2 + n^2 = 42k + 21$ for some integer k. We explore the following situations arising out from Eq. (4.1):

Table 4.1.

a	b	Remark
21	$n^2 / 42$	b is integer when n is an even integral multiple of 21.

42	$(n^2 + 21 \times 63) / 42$ $= (n^2 + 21) / 42 + 31$	b is integer when n is an odd integral multiple of 21.
63	$(n^2 + 42 \times 84) / 42$ $= n^2 / 42 + 84$	As for $a = 21$.
84	$(n^2 + 63 \times 105) / 42$ $= (n^2 + 21) / 42 + 157$	As for $a = 42$.
105	$(n^2 + 84 \times 126) / 42$ $= n^2 / 42 + 252$	As for $a = 21$.
126	$(n^2 + 105 \times 147) / 42$ $= (n^2 + 21) / 42 + 367$	As for $a = 42$.
147	$(n^2 + 126 \times 168) / 42$ $= n^2 / 42 + 504$	As for $a = 21$.
168	$(n^2 + 147 \times 189) / 42$ $= (n^2 + 21) / 42 + 661$	As for $a = 42$.
189	$(n^2 + 168 \times 210) / 42$ $= n^2 / 42 + 840$	As for $a = 21$.
210	$(n^2 + 189 \times 231) / 42$ $= (n^2 + 21) / 42 + 1039$	As for $a = 42$.

etc. Conclusively, b becomes integer when a takes *odd* integral multiples of 21 together with corresponding n as *even* integral multiples of 21 and vice-versa, i.e. when a assuming *even* but n taking *odd* integral multiples of 21. Thus, we have the following theorems. Symmetry of a and n in the sum $a^2 + n^2$ is also noticeable hence, the pairs (a, n) and (n, a) shall yield the same values of b.

Theorem 4.1. For $a = 21$, $b = n^2 / 42$ by Table 4.1. Hence, $n = 42$, 84, 126, 168, etc. yield the following identities.

n	b	$b + 21$	Identity	Ref. Chap.1
42	42	63	$21^2 + 42^2 + 42^2 = 63^2$	

			i.e. $1^2 + 2^2 + 2^2 = 3^2$	Th. 2.1
84	168	189	$21^2 + 84^2 + 168^2 = 189^2$	
			i.e. $1^2 + 4^2 + 8^2 = 9^2$	Th. 2.1
126	378	399	$21^2 + 126^2 + 378^2 = 399^2$	
			i.e. $1^2 + 6^2 + 18^2 = 19^2$	Th. 2.1
168	672	693	$21^2 + 168^2 + 672^2 = 693^2$	
			i.e. $1^2 + 8^2 + 32^2 = 33^2$	Th. 2.1
210	1050	1071	$21^2 + 210^2 + 1050^2 = 1071^2$	
			i.e. $1^2 + 10^2 + 50^2 = 51^2$	Th. 2.1
252	1512	1533	$21^2 + 252^2 + 1512^2 = 1533^2$	
			i.e. $1^2 + 12^2 + 72^2 = 73^2$	Th. 2.1
294	2058	2079	$21^2 + 294^2 + 2058^2 = 2079^2$	
			i.e. $1^2 + 14^2 + 98^2 = 99^2$	Th. 2.1
336	2688	2709	$21^2 + 336^2 + 2688^2 = 2709^2$	
			i.e. $1^2 + 16^2 + 128^2 = 129^2$	Th. 2.1
378	3402	3423	$21^2 + 378^2 + 3402^2 = 3423^2$	
			i.e. $1^2 + 18^2 + 162^2 = 163^2$	Th. 2.1
420	4200	4221	$21^2 + 420^2 + 4200^2 = 4221^2$	
			i.e. $1^2 + 20^2 + 200^2 = 201^2$	Th. 2.1

etc. //

Theorem 4.2. For $a = 42$, $b = (n^2 + 21)/42 + 31$ by Table 4.1. Hence, for $n = 63, 105, 147$ etc., we derive the following identities.

n	b	$b + 21$	Identity	Ref. Chap.1
63	126	147	$42^2 + 63^2 + 126^2 = 147^2$	
			i.e. $2^2 + 3^2 + 6^2 = 7^2$	Th. 2.2
105	294	315	$42^2 + 105^2 + 294^2 = 315^2$	
			i.e. $2^2 + 5^2 + 14^2 = 15^2$	Th. 2.2
147	546	567	$42^2 + 147^2 + 546^2 = 567^2$	
			i.e. $2^2 + 7^2 + 26^2 = 27^2$	Th. 2.2
189	882	903	$42^2 + 189^2 + 882^2 = 903^2$	
			i.e. $2^2 + 9^2 + 42^2 = 43^2$	Th. 2.2
231	1302	1323	$42^2 + 231^2 + 1302^2 = 1323^2$	
			i.e. $2^2 + 11^2 + 62^2 = 63^2$	Th. 2.2
273	1806	1827	$42^2 + 273^2 + 1806^2 = 1827^2$	
			i.e. $2^2 + 13^2 + 86^2 = 87^2$	Th. 2.2
315	2394	2415	$42^2 + 315^2 + 2394^2 = 2415^2$	
			i.e. $2^2 + 15^2 + 114^2 = 115^2$	Th. 2.2
357	3066	3087	$42^2 + 357^2 + 3066^2 = 3087^2$	
			i.e. $2^2 + 17^2 + 146^2 = 147^2$	Th. 2.2
399	3822	3843	$42^2 + 399^2 + 3822^2 = 3843^2$	
			i.e. $2^2 + 19^2 + 182^2 = 183^2$	Th. 2.2

etc. The choice of pair $(a, n) = (42, 21)$ is dropped as $(a, n) = (21, 42)$ yielding the same value of b is already considered in Theorem 4.1. //

Theorem 4.3. For $a = 63$, $b = n^2/42 + 84$ by Table 4.1. Hence, $n = 84, 126, 168, 210$, etc. yield the following identities.

n	b	$b + 21$	Identity	Ref. Chap.1
84	252	273	$63^2 + 84^2 + 252^2 = 273^2$	
			i.e. $3^2 + 4^2 + 12^2 = 13^2$	Th. 2.3
126	462	483	$63^2 + 126^2 + 462^2 = 483^2$	
			i.e. $3^2 + 6^2 + 22^2 = 23^2$	Th. 2.3
168	756	777	$63^2 + 168^2 + 756^2 = 777^2$	
			i.e. $3^2 + 8^2 + 36^2 = 37^2$	Th. 2.3
210	1134	1155	$63^2 + 210^2 + 1134^2 = 1155^2$	
			i.e. $3^2 + 10^2 + 54^2 = 55^2$	Th. 2.3
252	1596	1617	$63^2 + 252^2 + 1596^2 = 1617^2$	
			i.e. $3^2 + 12^2 + 76^2 = 77^2$	Th. 2.3
294	2142	2163	$63^2 + 294^2 + 2142^2 = 2163^2$	
			i.e. $3^2 + 14^2 + 102^2 = 103^2$	Th. 2.3
336	2772	2793	$63^2 + 336^2 + 2772^2 = 2793^2$	
			i.e. $3^2 + 16^2 + 132^2 = 133^2$	Th. 2.3
378	3486	3507	$63^2 + 378^2 + 3486^2 = 3507^2$	
			i.e. $3^2 + 18^2 + 166^2 = 167^2$	Th. 2.3
420	4284	4305	$63^2 + 420^2 + 4284^2 = 4305^2$	

			i.e. $3^2 + 20^2 + 204^2 = 205^2$	Th. 2.3

etc. The choice of pair $(a, n) = (63, 42)$ is dropped as $(a, n) = (42, 63)$ is already considered in Theorem 4.2. //

Theorem 4.4. For $a = 84$, $b = (n^2 + 21)/42 + 157$ by Table 4.1. Hence, $n = 105, 147, 189, 231, 273$, etc. yield the following identities.

n	b	$b + 19$	Identity	Ref. Chap.1
105	420	441	$84^2 + 105^2 + 420^2 = 441^2$	
			i.e. $4^2 + 5^2 + 20^2 = 21^2$	Th. 2.4
147	672	693	$84^2 + 147^2 + 672^2 = 693^2$	
			i.e. $4^2 + 7^2 + 32^2 = 33^2$	Th. 2.4
189	1008	1029	$84^2 + 189^2 + 1008^2 = 1029^2$	
			i.e. $4^2 + 9^2 + 48^2 = 49^2$	Th. 2.4
231	1428	1449	$84^2 + 231^2 + 1428^2 = 1449^2$	
			i.e. $4^2 + 11^2 + 68^2 = 69^2$	Th. 2.4
273	1932	1953	$84^2 + 273^2 + 1932^2 = 1953^2$	
			i.e. $4^2 + 13^2 + 92^2 = 93^2$	Th. 2.4
315	2520	2541	$84^2 + 315^2 + 2520^2 = 2541^2$	
			i.e. $4^2 + 15^2 + 120^2 = 121^2$	Th. 2.4
357	3192	3213	$84^2 + 357^2 + 3192^2 = 3213^2$	
			i.e. $4^2 + 17^2 + 152^2 = 153^2$	Th. 2.4

etc. The choices $(a, n) = (84, 21)$ and $(84, 63)$ are dropped as the corresponding results are already considered in Theorems 4.1 and 4.3 respectively. //

§ 5. General case: identities of the type

$$a^2 + n^2 + b^2 = (b + k)^2, \tag{5.1}$$

where k is an integer. Above type of identities require:

$$b = (a^2 + n^2 - k^2) / 2k = \{ (a - k).(a + k) + n^2 \} / 2k. \tag{5.2}$$

We examine the following choices of a and n:

(i) When both a, n are some integral multiples of k, say $a = a_1 k$ and $n = n_1 k$, where a_1, n_1 are any integers. Such choice of a, n reduces Eq. (5.2) to

$$b = (a_1^2 + n_1^2 - 1)\, k / 2 \Rightarrow b + k = (a_1^2 + n_1^2 + 1)\, k/2, \tag{5.3}$$

giving integral values of b and $b + k$ only when k is *even*, say $2k_1$, k_1 being an integer. As such, Eq. (5.3) further reduces to

$$b = (a_1^2 + n_1^2 - 1)\, k_1, \quad \text{and} \quad b + k = (a_1^2 + n_1^2 + 1)\, k_1. \tag{5.3b}$$

Thus, there exist identities of above type and we have the:

Theorem 5.1. For $a = 2a_1 k_1$ and $n = 2n_1 k_1$, there hold the identities

$$(2a_1)^2 + (2n_1)^2 + (a_1^2 + n_1^2 - 1)^2 = (a_1^2 + n_1^2 + 1)^2, \tag{5.4}$$

where the common factor k_1^2 is divided.

Note 6.1. Giving non-zero integral values to a_1, n_1, the identities discussed earlier for k taking even values 2, 4, 6, 8, etc., are deducible from the identity (5.4).

(ii) When both a, n are odd integral multiples of k, say $a = (2a_1 + 1)\, k$ and $n = (2n_1 + 1)\, k$, where a_1, n_1 are any integers. Such choices reduce Eq. (5.2) to

$$b = 2\,(a_1^2 + n_1^2 + a_1 + n_1)\, k + k/2, \tag{5.5}$$

which is integer only when k is even. Such case is already covered above in Part (i).

(iii) When both a, n are even integral multiples of k, say $a = 2a_1 k$ and $n = 2n_1 k$, where a_1, n_1 are any integers. Such choice reduces Eq. (5.2) to

$$b = 2\,(a_1^2 + n_1^2)\,k - k/2, \quad (5.6)$$

which is integer only when k is even. Such case is also already covered above in Part (i).

(iv) When one of a, n say a is an odd integral multiple of k, i.e. $a = (2a_1 + 1)\,k$ but $n = 2n_1 k$ is even multiple of k. This choice of a, n reduces Eq. (5.2) to

$$b = 2(a_1^2 + n_1^2 + a_1)\,k \quad (5.7) \Rightarrow b + k = \{2(a_1^2 + n_1^2 + a_1) + 1\}k, \quad (5.8)$$

giving integral values. Thus, there exist identities of above type and we have the:

Theorem 5.2. For $a = (2a_1 + 1)\,k$, $n = 2n_1 k$ there hold the identities:

$$(2a_1 + 1)^2 + (2n_1)^2 + \{2\,(a_1^2 + n_1^2 + a_1)\}^2 = \{2\,(a_1^2 + n_1^2 + a_1) + 1\}^2, \quad (5.9)$$

where the common factor k^2 is divided throughout.

Note 5.2. Giving non-zero integral values to a_1, n_1, the identities discussed earlier for k taking any integral value, are deducible from the identity (5.9).

CHAPTER 3

GENERALIZATIONS OF PYTHAGORAS THEOREM TO PENTAGONS - I

The previous two chapters dealt with the generalizations of Pythagoras theorem to quadrilateral ABCD composed of two right triangles ABC and ACD (cf. Fig. 1.1.1) with right angles at their vertices B and C. In other words, the solutions to the quadratic equation (1.1.1) in the set of (positive) integers were explored. Currently, we extend above result further for pentagons ABCDE with side-lengths a, b, c, d, e and having right angles at vertices B, C and D. However, before the main exercise we first re-visit Pythagoras theorem in some right triangles in the first section.

§ 1. Some triples satisfying Pythagoras theorem

1.1. As in Chapter 1, § 2, the integral values a, x, $x + 1$ will satisfy the Pythagoras theorem:

$$a^2 + x^2 = (x + 1)^2 = x^2 + 2x + 1,$$

if

$$x = (a^2 - 1) / 2. \qquad (1.1)$$

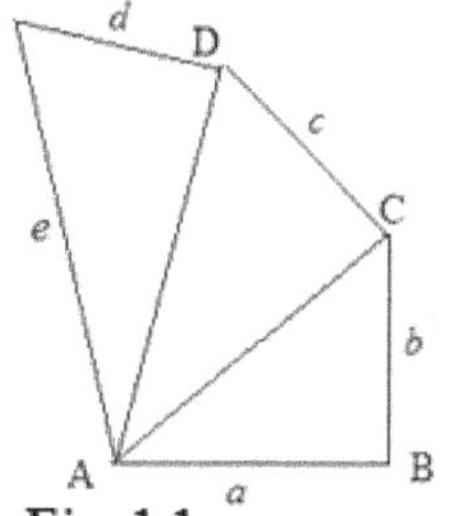

Fig. 1.1

is an integer. Thus, $a^2 - 1$ should be even for which a can have only odd values, say $2n + 1$, so that

$$x = \{(2n + 1)^2 - 1\} / 2 = 2n (n + 1),$$

where n is a whole number. Choices $n = 0$ or $a = 1$ make $x = 0$ and $x + 1 = 1$, as such the numbers a, x, $x + 1$ do not make a proper triangle and yield only trivial identity $1^2 + 0^2 = 1^2$. But, for other odd values of $a = 3$, 5, 7, 9, 11, etc. there exist triples

$$a = 2n + 1, \qquad x = 2n (n + 1) \quad \text{and} \quad x + 1 = 2n (n + 1) + 1,$$

satisfying the Pythagoras theorem:

$$(2n + 1)^2 + \{2n (n + 1)\}^2 \equiv \{2n (n + 1) + 1\}^2. \qquad (1.2)$$

Assigning different (integral) values to n in identity (1.2), one may easily compute the identities:

$$3^2 + 4^2 = 5^2,\ \ 5^2 + 12^2 = 13^2,\ \ 7^2 + 24^2 = 25^2,\ \ 9^2 + 40^2 = 41^2, \quad (1.3)$$

etc.

1.2. Similarly, the triples of integers of the type a, x, $x + 2$ satisfy Pythagoras theorem:

$$a^2 + x^2 = (x + 2)^2 = x^2 + 4x + 4,$$

if there exist integral values of x:

$$x = (a^2 - 4) / 4 \ = \ a^2 / 4 - 1,$$

for which a must be even, say $2n$. Accordingly, above equation yields

$$x \ = \ n^2 - 1 \qquad \text{and} \qquad x + 2 \ = \ n^2 + 1.$$

where n is a whole number. It may be noted that $n = 0$, forming the triple $a = 0$, $x = -1$ and $x + 2 = 1$, yields a trivial result $0^2 + (-1)^2 = 1^2$ only, and the numbers do not make a proper triangle. Also, $n = 1$, gives the triple $a = 2$, $x = 0$ and $x + 2 = 2$, which also yield a trivial result $2^2 + 0^2 = 2^2$. But, for the other (integral) values of $n = 2, 3, 4, 5$, etc. the triples $2n$, $n^2 - 1$ and $n^2 + 1$ satisfy the Pythagoras theorem:

$$(2n)^2 + (n^2 - 1)^2 \equiv \ (n^2 + 1)^2. \quad (1.4)$$

Giving different (integral) values to $n > 1$, above identity easily yields the identities:

$$4^2 + 3^2 = 5^2,\ \ 6^2 + 8^2 = 10^2,\ \ 8^2 + 15^2 = 17^2,\ \ 10^2 + 24^2 = 26^2, \quad (1.5)$$

etc.

1.3. Generalizing above approach, we explore for the triples a, x, $x + k$ satisfying the Pythagoras theorem:

$$a^2 + x^2 = (x + k)^2 = x^2 + 2kx + k^2,$$

if there exist any integral values of a and x for a given integer k:

$$x = (a^2 - k^2) / 2k \ = \ n \ \text{(say)} \qquad \Rightarrow \qquad a^2 = k\,(k + 2n),$$

for some whole number n. Accordingly, the triples

$$a = \sqrt{\{k\,(k+2n)\}}, \qquad x = n \quad \text{and} \quad x + k = n + k.$$

satisfy the identity:

$$k\,(k+2n) + n^2 \equiv (n+k)^2, \tag{1.6}$$

for some suitable choices of n. It may be noted that $n = 0$, giving the triples $a = k$, $x = 0$ and $x + k = k$, yields a trivial result $k^2 + 0^2 = k^2$ only and there exists no proper triangle. Also, $n = 1$ does not make $a = \sqrt{\{k.(k+2)\}}$ integer for any $k = 1, 2, 3, \ldots$

It may be noted that, for $k = 1$, one can derive the identities (1.3) discussed in Sub-section 1.1 from the general formula (1.6) for special choices of $n = 4, 12, 24, 40, \ldots$ Also, for $k = 2$, the identities (1.5) in Sub-section 1.2 can be derived from the general result (1.6) for $n = 3, 8, 15, 24, \ldots$

Following the methods employed in Chapter 1, we proceed similarly for exploration of generalizations of Pythagoras theorem to pentagons.

§ 2. Identities of the type $a^2 + b^2 + c^2 + x^2 = (x+1)^2$

2.1. Expanding the right hand member, and dropping the common terms one easily derives the value of x in terms of a, b and c in order to satisfy above identity:

$$x = (a^2 + b^2 + c^2 - 1) / 2, \tag{2.1}$$

which takes integral values when the sum $a^2 + b^2 + c^2$ gets odd value. Thus, there arise two choices of the integers a, b, c:

(i) Either all taking odd values, say $a = 2p + 1$, $b = 2q + 1$, $c = 2r + 1$ making x also odd:

$$x = 2\,(p^2 + q^2 + r^2 + p + q + r) + 1, \tag{2.2}$$

(ii) Or, any two even, say $a = 2p$, $b = 2q$, but $c = 2r + 1$ making x even:

$$x = 2\,(p^2 + q^2 + r^2 + r). \tag{2.3}$$

Giving different integral values to p, q, r there exist two different types of quadruples: all a, b, c, x taking odd values, and (alternately) a, b, x even but c odd. Accordingly, there exist identities of above types. For instance, the choice $p = q = r = 0$, giving all $a = b = c = x = 1$, yields the identity:

$$1^2 + 1^2 + 1^2 + 1^2 = 2^2. \quad (2.4)$$

On the other hand, taking $p = q = r = 1$, we get two different types of identities from above diverse cases:

(i) $a = b = c = 3$, $x = 13$ yielding the identity:

$$3^2 + 3^2 + 3^2 + 13^2 = 14^2; \quad (2.5)$$

and

(ii) $a = b = 2$, $c = 3$, $x = 8$ giving the identity:

$$2^2 + 2^2 + 3^2 + 8^2 = 9^2. \quad (2.6)$$

This way, one can derive a large number of identities by taking different values of p, q, r.

2.2. Since we have derived a large number of identities representing generalizations of Pythagoras theorem to quadrilaterals in our previous chapters, we follow entirely different approach to compute the additional (fourth) number x according to Eq. (2.1), taking into account the sum $a^2 + b^2 + c^2$ as the perfect square: $(c + 1)^2$. Thus,

$$x = \{(c + 1)^2 - 1\}/2 = c\,(c/2 + 1). \quad (2.7)$$

Extending the results of Chapter 1, we have the following theorems.

Theorem 2.1. For $a = 1$, there hold the identities:

b	c	$x = c\,(c/2 + 1)$	Identity for pentagons
2	2	4	$1^2 + 2^2 + 2^2 + 4^2 = 5^2$
4	8	40	$1^2 + 4^2 + 8^2 + 40^2 = 41^2$
6	18	180	$1^2 + 6^2 + 18^2 + 180^2 = 181^2$
8	32	544	$1^2 + 8^2 + 32^2 + 544^2 = 545^2$
10	50	1300	$1^2 + 10^2 + 50^2 + 1300^2 = 1301^2$

12	72	2664	$1^2 + 12^2 + 72^2 + 2664^2 = 2665^2$

etc. //

Theorem 2.2. For $a = 2$ and Eq. (2.7), here hold the identities:

b	c	x	Identity for pentagons	Reference
1	2	4	$2^2 + 1^2 + 2^2 + 4^2 = 5^2$	Th. 2.1
3	6	24	$2^2 + 3^2 + 6^2 + 24^2 = 25^2$	
5	14	112	$2^2 + 5^2 + 14^2 + 112^2 = 113^2$	
7	26	364	$2^2 + 7^2 + 26^2 + 364^2 = 365^2$	
9	42	924	$2^2 + 9^2 + 42^2 + 924^2 = 925^2$	
11	62	1984	$2^2 + 11^2 + 62^2 + 1984^2 = 1985^2$	
13	86	3784	$2^2 + 13^2 + 86^2 + 3784^2 = 3785^2$	

etc. //

Theorem 2.3. For $a = 3$ and Eq. (2.7), there hold the identities:

b	c	x	Identity for pentagons	Reference
2	6	24	$3^2 + 2^2 + 6^2 + 24^2 = 25^2$	Th. 2.2
4	12	84	$3^2 + 4^2 + 12^2 + 84^2 = 85^2$	
6	22	264	$3^2 + 6^2 + 22^2 + 264^2 = 265^2$	
8	36	684	$3^2 + 8^2 + 36^2 + 684^2 = 685^2$	
10	54	1512	$3^2 + 10^2 + 54^2 + 1512^2 = 1513^2$	
12	76	2964	$3^2 + 12^2 + 76^2 + 2964^2 = 2965^2$	

etc. //

Theorem 2.4. For $a = 4$ and Eq. (2.7), there hold the identities:

b	c	x	Identity for pentagons	Reference
1	8	40	$4^2 + 1^2 + 8^2 + 40^2 = 41^2$	Th. 2.1
3	12	84	$4^2 + 3^2 + 12^2 + 84^2 = 85^2$	Th. 2.3
5	20	220	$4^2 + 5^2 + 20^2 + 220^2 = 221^2$	
7	32	544	$4^2 + 7^2 + 32^2 + 544^2 = 545^2$	
9	48	1200	$4^2 + 9^2 + 48^2 + 1200^2 = 1201^2$	
11	68	2380	$4^2 + 11^2 + 68^2 + 2380^2 = 2381^2$	

etc. //

Theorem 2.5. For $a = 5$ and Eq. (2.7), here hold the identities:

b	c	x	Identity for pentagons	Ref.
2	14	112	$5^2 + 2^2 + 14^2 + 112^2 = 113^2$	Th. 2.2
4	20	220	$5^2 + 4^2 + 20^2 + 220^2 = 221^2$	Th. 2.4
6	30	480	$5^2 + 6^2 + 30^2 + 480^2 = 481^2$	
8	44	1012	$5^2 + 8^2 + 44^2 + 1012^2 = 1013^2$	
10	62	1984	$5^2 + 10^2 + 62^2 + 1984^2 = 1985^2$	
12	84	3612	$5^2 + 12^2 + 84^2 + 3612^2 = 3613^2$	

etc. //

Theorem 2.6. For $a = 6$ and Eq. (2.7), there hold the identities:

b	c	x	Identity for pentagons	Ref.
1	18	180	$6^2 + 1^2 + 18^2 + 180^2 = 181^2$	Th. 2.1
3	22	264	$6^2 + 3^2 + 22^2 + 264^2 = 265^2$	Th. 2.3
5	30	480	$6^2 + 5^2 + 30^2 + 480^2 = 481^2$	Th. 2.5
7	42	924	$6^2 + 7^2 + 42^2 + 924^2 = 925^2$	
9	58	1740	$6^2 + 9^2 + 58^2 + 1740^2 = 1741^2$	
11	78	3120	$6^2 + 11^2 + 78^2 + 3120^2 = 3121^2$	

etc. //

Theorem 2.7. For $a = 7$ and Eq. (2.7), there hold the identities:

b	c	x	Identity for pentagons	Ref.
2	26	364	$7^2 + 2^2 + 26^2 + 364^2 = 365^2$	Th. 2.2
4	32	544	$7^2 + 4^2 + 32^2 + 544^2 = 545^2$	Th. 2.4
6	42	924	$7^2 + 6^2 + 42^2 + 924^2 = 925^2$	Th. 2.6
8	56	1624	$7^2 + 8^2 + 56^2 + 1624^2 = 1625^2$	
10	74	2812	$7^2 + 10^2 + 74^2 + 2812^2 = 2813^2$	
12	96	4704	$7^2 + 12^2 + 96^2 + 4704^2 = 4705^2$	

etc. //

Theorem 2.8. For $a = 8$ and Eq. (2.7), there hold the identities:

b	c	x	Identity for pentagons	Ref.
1	32	544	$8^2 + 1^2 + 32^2 + 544^2 = 545^2$	Th. 2.1

3	36	684	$8^2 + 3^2 + 36^2 + 684^2 = 685^2$	Th. 2.3
5	44	1012	$8^2 + 5^2 + 44^2 + 1012^2 = 1013^2$	Th. 2.5
7	56	1624	$8^2 + 7^2 + 56^2 + 1624^2 = 1625^2$	Th. 2.7
9	72	2664	$8^2 + 9^2 + 72^2 + 2664^2 = 2665^2$	
11	92	4324	$8^2 + 11^2 + 92^2 + 4324^2 = 4325^2$	

etc. //

Theorem 2.9. For $a = 9$ and Eq. (2.7), there hold the identities:

b	c	x	Identity for pentagons	Ref.
2	42	924	$9^2 + 2^2 + 42^2 + 924^2 = 925^2$	Th. 2.2
4	48	1200	$9^2 + 4^2 + 48^2 + 1200^2 = 1201^2$	Th. 2.4
6	58	1740	$9^2 + 6^2 + 58^2 + 1740^2 = 1741^2$	Th. 2.6
8	72	2664	$9^2 + 8^2 + 72^2 + 2664^2 = 2665^2$	Th. 2.8
10	90	4140	$9^2 + 10^2 + 90^2 + 4140^2 = 4141^2$	
12	112	6384	$9^2 + 12^2 + 112^2 + 6384^2 = 6385^2$	

etc. //

Theorem 2.10. For $a = 10$ and Eq. (2.7), there hold the identities:

b	c	x	Identity for pentagons	Ref.
1	50	1300	$10^2 + 1^2 + 50^2 + 1300^2 = 1301^2$	Th. 2.1
3	54	1512	$10^2 + 3^2 + 54^2 + 1512^2 = 1513^2$	Th. 2.3
5	62	1984	$10^2 + 5^2 + 62^2 + 1984^2 = 1985^2$	Th. 2.5
7	74	2812	$10^2 + 7^2 + 74^2 + 2812^2 = 2813^2$	Th. 2.7

9	90	4140	$10^2 + 9^2 + 90^2 + 4140^2 = 4141^2$	Th. 2.9
11	110	6160	$10^2 + 11^2 + 110^2 + 6160^2 = 6161^2$	

etc. //

§ 3. Identities of the type $a^2 + b^2 + c^2 + x^2 = (x + 2)^2$

Above type of identities require:

$$x = (a^2 + b^2 + c^2 - 4)/4 = (a^2 + b^2 + c^2)/4 - 1, \qquad (3.1)$$

where a , b, c assume some suitable integral values making x integer. For $a = 1$, Eq. (3.1) yields $x = (b^2 + c^2 - 3)/4$, which cannot assume integral values unless $k \equiv b^2 + c^2 - 3$ becomes an integral multiple of 4. There arise three types of choices for b, c: both even say $2m$, $2n$; both odd $2m + 1$, $2n + 1$ and one of them even, say $b = 2m$, but c odd, say $2n + 1$. Accordingly, x assumes the values

$$m^2 + n^2 - 3/4, \qquad m^2 + n^2 + m + n - 1/4, \qquad m^2 + n^2 + n - 1/2.$$

None of these yield whole numbers for integers m, n. Hence, there exist no such identities for $a = 1$.

But, for $a = 2$, Eq. (3.1) determines

$$x = (b^2 + c^2) / 4 \qquad (3.2)$$

assuming integral values when both b, c are even. It can be verified, as above, x does not assume integral values for other choices of b, c: both taking odd or one odd while the other even. Hence, there exist identities for $a - 2$. In the following, we compute x for different integral values of $a = 2, 3, 4, 5$, etc.

a	x	Remark
2	$(b^2 + c^2)/4$	There exist identities.
3	$(b^2 + c^2 + 5)/4 = (b^2 + c^2 + 1)/4 + 1$	As for $a = 1$.
4	$(b^2 + c^2 + 12)/4 = (b^2 + c^2)/4 + 3$	As for $a = 2$.

5	$(b^2 + c^2 + 21)/4 = (b^2 + c^2 + 1)/4 + 5$	As for $a = 1$.
6	$(b^2 + c^2 + 32)/4 = (b^2 + c^2)/4 + 8$	As for $a = 2$.
7	$(b^2 + c^2 + 45)/4 = (b^2 + c^2 + 1)/4 + 11$	As for $a = 1$.
8	$(b^2 + c^2 + 60)/4 = (b^2 + c^2)/4 + 15$	As for $a = 2$.
9	$(b^2 + c^2 + 77)/4 = (b^2 + c^2 + 1)/4 + 19$	As for $a = 1$.
10	$(b^2 + c^2 + 96)/4 = (b^2 + c^2)/4 + 24$	As for $a = 2$.

etc. Conclusively, there exist identities for even values of a, but no identities for odd values of a.

Note 3.1. Because of symmetry of a, b in the value of x given by Eq. (3.1), the results will be same when they take interchanging values, i.e. $(a, b) = (b, a)$, when a, b assume values 2, 4, 6, 8, 10, 12, …

Theorem 3.1. For $a = b = 2$ and Eq. (3.2), there hold the identities:

c	x	Identity for pentagons	Equivalently	Ref.
2	2	$2^2 + 2^2 + 2^2 + 2^2 = 4^2$	$1^2 + 1^2 + 1^2 + 1^2 = 2^2$	Eq. (2.4)
4	5	$2^2 + 2^2 + 4^2 + 5^2 = 7^2$		
6	10	$2^2 + 2^2 + 6^2 + 10^2 = 12^2$	$1^2 + 1^2 + 3^2 + 5^2 = 6^2$	
8	17	$2^2 + 2^2 + 8^2 + 17^2 = 19^2$		
10	26	$2^2 + 2^2 + 10^2 + 26^2 = 28^2$	$1^2 + 1^2 + 5^2 + 13^2 = 14^2$	
12	37	$2^2 + 2^2 + 12^2 + 37^2 = 39^2$		

etc. //

Theorem 3.2. For $(a, b) = (2, 4)$ or $(4, 2)$ and Eq. (3.2), there hold the identities:

c	x	Identity for pentagons	Equivalently	Ref.
2	5	$2^2 + 4^2 + 2^2 + 5^2 = 7^2$		Th. 3.1
4	8	$2^2 + 4^2 + 4^2 + 8^2 = 10^2$	$1^2 + 2^2 + 2^2 + 4^2 = 5^2$	Th. 2.1
6	13	$2^2 + 4^2 + 6^2 + 13^2 = 15^2$		
8	20	$2^2 + 4^2 + 8^2 + 20^2 = 22^2$	$1^2 + 2^2 + 4^2 + 10^2 = 11^2$	
10	29	$2^2 + 4^2 + 10^2 + 29^2 = 31^2$		
12	40	$2^2 + 4^2 + 12^2 + 40^2 = 42^2$	$1^2 + 2^2 + 6^2 + 20^2 = 21^2$	

etc. //

Theorem 3.3. For $(a, b) = (2, 6)$ or $(6, 2)$ and Eq. (3.2), there hold the identities:

c	x	Identity for pentagons	Equivalently	Ref.
2	10	$2^2 + 6^2 + 2^2 + 10^2 = 12^2$	$1^2 + 3^2 + 1^2 + 5^2 = 6^2$	Th. 3.1
4	13	$2^2 + 6^2 + 4^2 + 13^2 = 15^2$		Th. 3.2
6	18	$2^2 + 6^2 + 6^2 + 18^2 = 20^2$	$1^2 + 3^2 + 3^2 + 9^2 = 10^2$	
8	25	$2^2 + 6^2 + 8^2 + 25^2 = 27^2$		
10	34	$2^2 + 6^2 + 10^2 + 34^2 = 36^2$	$1^2 + 3^2 + 5^2 + 17^2 = 18^2$	
12	45	$2^2 + 6^2 + 12^2 + 45^2 = 47^2$		

etc. //

Theorem 3.4. For $(a, b) = (2, 8)$ or $(8, 2)$ and Eq. (3.2), there hold the identities:

c	x	Identity for pentagons	Equivalently	Ref.
2	17	$2^2 + 8^2 + 2^2 + 17^2 = 19^2$		Th. 3.1
4	20	$2^2 + 8^2 + 4^2 + 20^2 = 22^2$	$1^2 + 4^2 + 2^2 + 10^2 = 11^2$	Th. 3.2
6	25	$2^2 + 8^2 + 6^2 + 25^2 = 27^2$		Th. 3.3
8	32	$2^2 + 8^2 + 8^2 + 32^2 = 34^2$	$1^2 + 4^2 + 4^2 + 16^2 = 17^2$	
10	41	$2^2 + 8^2 + 10^2 + 41^2 = 43^2$		
12	52	$2^2 + 8^2 + 12^2 + 52^2 = 54^2$	$1^2 + 4^2 + 6^2 + 26^2 = 27^2$	

etc. //

Theorem 3.5. For $(a, b) = (2, 10)$ or $(10, 2)$ and Eq. (3.2), there hold the identities:

c	x	Identity for pentagons	Equivalently	Ref.
2	26	$2^2 + 10^2 + 2^2 + 26^2 = 28^2$	$1^2 + 5^2 + 1^2 + 13^2 = 14^2$	Th. 3.1
4	29	$2^2 + 10^2 + 4^2 + 29^2 = 31^2$		Th. 3.2
6	34	$2^2 + 10^2 + 6^2 + 34^2 = 36^2$	$1^2 + 5^2 + 3^2 + 17^2 = 18^2$	Th. 3.3
8	41	$2^2 + 10^2 + 8^2 + 41^2 = 43^2$		Th. 3.4
10	50	$2^2 + 10^2 + 10^2 + 50^2 = 52^2$	$1^2 + 5^2 + 5^2 + 25^2 = 26^2$	
12	61	$2^2 + 10^2 + 12^2 + 61^2 = 63^2$		

etc. //

Theorem 3.6. For $a = b = 4$, so that Eq. (3.1) $\Rightarrow x = c^2/4 + 7$, there hold the following identities:

c	x	Identity	Equivalently	Ref.
2	8	$4^2 + 4^2 + 2^2 + 8^2 = 10^2$	$2^2 + 2^2 + 1^2 + 4^2 = 5^2$	Th. 2.1
4	11	$4^2 + 4^2 + 4^2 + 11^2 = 13^2$		
6	16	$4^2 + 4^2 + 6^2 + 16^2 = 18^2$	$2^2 + 2^2 + 3^2 + 8^2 = 9^2$	Eq. (2.6)
8	23	$4^2 + 4^2 + 8^2 + 23^2 = 25^2$		
10	32	$4^2 + 4^2 + 10^2 + 32^2 = 34^2$	$2^2 + 2^2 + 5^2 + 16^2 = 17^2$	
12	43	$4^2 + 4^2 + 12^2 + 43^2 = 45^2$		

etc. //

Theorem 3.7. For $(a, b) = (4, 6)$ or $(6, 4)$, so that Eq. (3.1) $\Rightarrow x = c^2/4 + 12$, there hold the following identities:

c	x	Identity	Equivalently	Ref.
2	13	$4^2 + 6^2 + 2^2 + 13^2 = 15^2$		Th. 3.2
4	16	$4^2 + 6^2 + 4^2 + 16^2 = 18^2$	$2^2 + 3^2 + 2^2 + 8^2 = 9^2$	Eq. (2.6)
6	21	$4^2 + 6^2 + 6^2 + 21^2 = 23^2$		
8	28	$4^2 + 6^2 + 8^2 + 28^2 = 30^2$	$2^2 + 3^2 + 4^2 + 14^2 = 15^2$	
10	37	$4^2 + 6^2 + 10^2 + 37^2 = 39^2$		
12	48	$4^2 + 6^2 + 12^2 + 48^2 = 50^2$	$2^2 + 3^2 + 6^2 + 24^2 = 25^2$	Th. 2.2

etc. //

Theorem 3.8. For $(a, b) = (4, 8)$ or $(8, 4)$, so that Eq. (3.1) $\Rightarrow x = c^2/4 + 19$, there hold the following identities:

c	x	Identity	Equivalently	Ref.
2	20	$4^2 + 8^2 + 2^2 + 20^2 = 22^2$	$2^2 + 4^2 + 1^2 + 10^2 = 11^2$	Th. 3.2
4	23	$4^2 + 8^2 + 4^2 + 23^2 = 25^2$		Th. 3.6
6	28	$4^2 + 8^2 + 6^2 + 28^2 = 30^2$	$2^2 + 4^2 + 3^2 + 14^2 = 15^2$	Th. 3.7
8	35	$4^2 + 8^2 + 8^2 + 35^2 = 37^2$		
10	44	$4^2 + 8^2 + 10^2 + 44^2 = 46^2$	$2^2 + 4^2 + 5^2 + 22^2 = 23^2$	
12	55	$4^2 + 8^2 + 12^2 + 55^2 = 57^2$		

etc. //

Theorem 3.9. For $(a, b) = (4, 10)$ or $(10, 4)$, so that Eq. (3.1) $\Rightarrow x = c^2/4 + 28$, there hold the following identities:

c	x	Identity	Equivalently	Ref.
2	29	$4^2 + 10^2 + 2^2 + 29^2 = 31^2$		Th. 3.2
4	32	$4^2 + 10^2 + 4^2 + 32^2 = 34^2$	$2^2 + 5^2 + 2^2 + 16^2 = 17^2$	Th. 3.6
6	37	$4^2 + 10^2 + 6^2 + 37^2 = 39^2$		Th. 3.7
8	44	$4^2 + 10^2 + 8^2 + 44^2 = 46^2$	$2^2 + 5^2 + 4^2 + 22^2 = 23^2$	Th. 3.8
10	53	$4^2 + 10^2 + 10^2 + 53^2 = 55^2$		
12	64	$4^2 + 10^2 + 12^2 + 64^2 = 66^2$	$2^2 + 5^2 + 6^2 + 32^2 = 33^2$	

etc. //

Similarly, we derive the results for $a = 6$, so that Eq. (3.1) $\Rightarrow x = (b^2 + c^2)/4 + 8$ and have the following theorems.

Theorem 3.10. For $a = b = 6$, so that Eq. (3.1) $\Rightarrow x = c^2/4 + 17$, there hold the following identities:

c	x	Identity	Equivalently	Ref.
2	18	$6^2 + 6^2 + 2^2 + 18^2 = 20^2$	$3^2 + 3^2 + 1^2 + 9^2 = 10^2$	Th. 3.3
4	21	$6^2 + 6^2 + 4^2 + 21^2 = 23^2$		Th. 3.7
6	26	$6^2 + 6^2 + 6^2 + 26^2 = 28^2$	$3^2 + 3^2 + 3^2 + 13^2 = 14^2$	Eq. (2.5)
8	33	$6^2 + 6^2 + 8^2 + 33^2 = 35^2$		
10	42	$6^2 + 6^2 + 10^2 + 42^2 = 44^2$	$3^2 + 3^2 + 5^2 + 21^2 = 22^2$	
12	53	$6^2 + 6^2 + 12^2 + 53^2 = 55^2$		

etc. //

Theorem 3.11. For $(a, b) = (6, 8)$ or $(8, 6)$, so that Eq. (3.1) $\Rightarrow x = c^2/4 + 24$, there hold the following identities:

c	x	Identity	Equivalently	Ref.
2	25	$6^2 + 8^2 + 2^2 + 25^2 = 27^2$		Th. 3.3
4	28	$6^2 + 8^2 + 4^2 + 28^2 = 30^2$	$3^2 + 4^2 + 2^2 + 14^2 = 15^2$	Th. 3.7
6	33	$6^2 + 8^2 + 6^2 + 33^2 = 35^2$		Th. 3.10
8	40	$6^2 + 8^2 + 8^2 + 40^2 = 42^2$	$3^2 + 4^2 + 4^2 + 20^2 = 21^2$	
10	49	$6^2 + 8^2 + 10^2 + 49^2 = 51^2$		
12	60	$6^2 + 8^2 + 12^2 + 60^2 = 62^2$	$3^2 + 4^2 + 6^2 + 30^2 = 31^2$	

etc. //

Theorem 3.12. For $(a, b) = (6, 10)$ or $(10, 6)$, so that Eq. (3.1) $\Rightarrow x = c^2/4 + 33$, there hold the following identities:

c	x	Identity	Equivalently	Ref.
2	34	$6^2 + 10^2 + 2^2 + 34^2 = 36^2$	$3^2 + 5^2 + 1^2 + 17^2 = 18^2$	Th. 3.3
4	37	$6^2 + 10^2 + 4^2 + 37^2 = 39^2$		Th. 3.7
6	42	$6^2 + 10^2 + 6^2 + 42^2 = 44^2$	$3^2 + 5^2 + 3^2 + 21^2 = 22^2$	Th. 3.10
8	49	$6^2 + 10^2 + 8^2 + 49^2 = 51^2$		Th. 3.11
10	58	$6^2 + 10^2 + 10^2 + 58^2 = 60^2$	$3^2 + 5^2 + 5^2 + 29^2 = 30^2$	
12	69	$6^2 + 10^2 + 12^2 + 69^2 = 71^2$		

etc. //

Similarly, we derive the results for $a = 8$, so that Eq. (3.1) $\Rightarrow x = (b^2 + c^2)/4 + 15$ and have the following theorems.

Theorem 3.13. For $a = b = 8$, so that Eq. (3.1) $\Rightarrow x = c^2/4 + 31$, there hold the following identities:

c	x	Identity	Equivalently	Ref.
2	32	$8^2 + 8^2 + 2^2 + 32^2 = 34^2$	$4^2 + 4^2 + 1^2 + 16^2 = 17^2$	Th. 3.4
4	35	$8^2 + 8^2 + 4^2 + 35^2 = 37^2$		Th. 3.8
6	40	$8^2 + 8^2 + 6^2 + 40^2 = 42^2$	$4^2 + 4^2 + 3^2 + 20^2 = 21^2$	Th. 3.11
8	47	$8^2 + 8^2 + 8^2 + 47^2 = 49^2$		
10	56	$8^2 + 8^2 + 10^2 + 56^2 = 58^2$	$4^2 + 4^2 + 5^2 + 28^2 = 29^2$	
12	67	$8^2 + 8^2 + 12^2 + 67^2 = 69^2$		

etc. //

Theorem 3.14. For $(a, b) = (8, 10)$ or $(10, 8)$, so that Eq. (3.1) $\Rightarrow x = c^2/4 + 40$, there hold the following identities:

c	x	Identity	Equivalently	Ref.
2	41	$8^2 + 10^2 + 2^2 + 41^2 = 43^2$		Th. 3.4
4	44	$8^2 + 10^2 + 4^2 + 44^2 = 46^2$	$4^2 + 5^2 + 2^2 + 22^2 = 23^2$	Th. 3.8
6	49	$8^2 + 10^2 + 6^2 + 49^2 = 51^2$		Th. 3.11
8	56	$8^2 + 10^2 + 8^2 + 56^2 = 58^2$	$4^2 + 5^2 + 4^2 + 28^2 = 29^2$	Th. 3.13
10	65	$8^2 + 10^2 + 10^2 + 65^2 = 67^2$		
12	76	$8^2 + 10^2 + 12^2 + 76^2 = 78^2$	$4^2 + 5^2 + 6^2 + 38^2 = 39^2$	

etc. //

Similarly, we derive the results for $a = 10$, so that Eq. (3.1) $\Rightarrow x = (b^2 + c^2)/4 + 24$ and have the following theorems.

Theorem 3.15. For $a = b = 10$, so that Eq. (3.1) $\Rightarrow x = c^2/4 + 49$, there hold the following identities:

c	x	Identity	Equivalently	Ref.
2	50	$10^2 + 10^2 + 2^2 + 50^2 = 52^2$	$5^2 + 5^2 + 1^2 + 25^2 = 26^2$	Th. 3.5
4	53	$10^2 + 10^2 + 4^2 + 53^2 = 55^2$		Th. 3.9
6	58	$10^2 + 10^2 + 6^2 + 58^2 = 60^2$	$5^2 + 5^2 + 3^2 + 29^2 = 30^2$	Th. 3.12
8	65	$10^2 + 10^2 + 8^2 + 65^2 = 67^2$		Th. 3.14
10	74	$10^2 + 10^2 + 10^2 + 74^2 = 76^2$	$5^2 + 5^2 + 5^2 + 37^2 = 38^2$	
12	85	$10^2 + 10^2 + 12^2 + 85^2 = 87^2$		

etc. //

§ 4. Identities of the type $a^2 + b^2 + c^2 + x^2 = (x + 3)^2$

Above type of identities require:

$$x = (a^2 + b^2 + c^2 - 9)/6, \quad (4.1)$$

where a, b and c assume some suitable integral values making x integer. Following the line of approach as explained in Sub-section 2.2, to compute x according to Eq. (4.1), we proceed as follows. Taking into account the sum $a^2 + b^2 + c^2 \equiv (c + 3)^2$ as computed in [3], we get

$$x = \{(c+3)^2 - 9\}/6 = c\,(c/6 + 1). \quad (4.2)$$

Thus, extending the results of the Chapter 1, we have the following:

Theorem 4.1. For $a = 3$ and Eq. (4.2), there hold the identities:

b	c	x	Identity	Equivalently	Ref.
6	6	12	$3^2 + 6^2 + 6^2 + 12^2 = 15^2$	$1^2 + 2^2 + 2^2 + 4^2 = 5^2$	Th. 2.1
12	24	120	$3^2 + 12^2 + 24^2 + 120^2 = 123^2$	$1^2 + 4^2 + 8^2 + 40^2 = 41^2$	”
18	54	540	$3^2 + 18^2 + 54^2 + 540^2 = 543^2$	$1^2 + 6^2 + 18^2 + 180^2 = 181^2$	”
24	96	1632	$3^2 + 24^2 + 96^2 + 1632^2 = 1635^2$	$1^2 + 8^2 + 32^2 + 544^2 = 545^2$	”
30	150	3900	$3^2 + 30^2 + 150^2\ 3900^2 = 3903^2$	$1^2 + 10^2 + 50^2 + 1300^2 = 1301^2$	”
36	216	7992	$3^2 + 36^2 + 216^2 + 7992^2 = 7995^2$	$1^2 + 12^2 + 72^2 + 2664^2 = 2665^2$	”

etc. //

Theorem 4.2. For $a = 6$ and Eq. (4.2), there hold the identities:

b	c	x	Identity	Equivalently	Ref.
3	6	12	$6^2 + 3^2 + 6^2 + 12^2 = 15^2$	$2^2 + 1^2 + 2^2 + 4^2 = 5^2$	Th. 2.1

9	18	72	$6^2 + 9^2 + 18^2 + 72^2 = 75^2$	$2^2 + 3^2 + 6^2 + 24^2 = 25^2$	Th. 2.2
15	42	336	$6^2 + 15^2 + 42^2 + 336^2 = 339^2$	$2^2 + 5^2 + 14^2 + 112^2 = 113^2$	”
21	78	1092	$6^2 + 21^2 + 78^2 + 1092^2 = 1095^2$	$2^2 + 7^2 + 26^2 + 364^2 = 365^2$	”
27	126	2772	$6^2 + 27^2 + 126^2 + 2772^2 = 2775^2$	$2^2 + 9^2 + 42^2 + 924^2 = 925^2$	”
33	186	5952	$6^2 + 33^2 + 186^2 + 5952^2 = 5955^2$	$2^2 + 11^2 + 62^2 + 1984^2 = 1985^2$	”

etc. //

Theorem 4.3. For $a = 9$ and Eq. (4.2), there hold the identities:

b	c	x	Identity	Equivalently	Ref.
6	18	72	$9^2 + 6^2 + 18^2 + 72^2 = 75^2$	$3^2 + 2^2 + 6^2 + 24^2 = 25^2$	Th. 2.2
12	36	252	$9^2 + 12^2 + 36^2 + 252^2 = 255^2$	$3^2 + 4^2 + 12^2 + 84^2 = 85^2$	Th. 2.3
18	66	792	$9^2 + 18^2 + 66^2 + 792^2 = 795^2$	$3^2 + 6^2 + 22^2 + 264^2 = 265^2$	”
24	108	2052	$9^2 + 24^2 + 108^2 + 2052^2 = 2055^2$	$3^2 + 8^2 + 36^2 + 684^2 = 685^2$	”
30	162	4536	$9^2 + 30^2 + 162^2 + 4536^2 = 4539^2$	$3^2 + 10^2 + 54^2 + 1512^2 = 1513^2$	”
36	228	8892	$9^2 + 36^2 + 228^2 + 8892^2 = 8895^2$	$3^2 + 12^2 + 76^2 + 2964^2 = 2965^2$	”

etc. //

Theorem 4.4. For $a = 12$ and Eq. (4.2), there hold the identities:

b	c	x	Identity	Equivalently	Ref.
3	24	120	$12^2 + 3^2 + 24^2 + 120^2 = 123^2$	$4^2 + 1^2 + 8^2 + 40^2 = 41^2$	Th. 2.1

9	36	252	$12^2 + 9^2 + 36^2 + 252^2 = 255^2$	$4^2 + 3^2 + 12^2 + 84^2 = 85^2$	Th. 2.3
15	60	660	$12^2 + 15^2 + 60^2 + 660^2 = 663^2$	$4^2 + 5^2 + 20^2 + 220^2 = 221^2$	Th. 2.4
21	96	1632	$12^2 + 21^2 + 96^2 + 1632^2 = 1635^2$	$4^2 + 7^2 + 32^2 + 544^2 = 545^2$	”
27	144	3600	$12^2 + 27^2 + 144^2 + 3600^2 = 3603^2$	$4^2 + 9^2 + 48^2 + 1200^2 = 1201^2$	”
33	204	7140	$12^2 + 33^2 + 204^2 + 7140^2 = 7143^2$	$4^2 + 11^2 + 68^2 + 2380^2 = 2381^2$	”

etc. //

Theorem 4.5. For $a = 15$ and Eq. (4.2), there hold the identities

b	c	x	Identity	Equivalently	Ref.
6	42	336	$15^2 + 6^2 + 42^2 + 336^2 = 339^2$	$5^2 + 2^2 + 14^2 + 112^2 = 113^2$	Th. 2.2
12	60	660	$15^2 + 12^2 + 60^2 + 660^2 = 663^2$	$5^2 + 4^2 + 20^2 + 220^2 = 221^2$	Th. 2.4
18	90	1440	$15^2 + 18^2 + 90^2 + 1440^2 = 1443^2$	$5^2 + 6^2 + 30^2 + 480^2 = 481^2$	Th. 2.5
24	132	3036	$15^2 + 24^2 + 132^2 + 3036^2 = 3039^2$	$5^2 + 8^2 + 44^2 + 1012^2 = 1013^2$	”
30	186	5952	$15^2 + 30^2 + 186^2 + 5952^2 = 5955^2$	$5^2 + 10^2 + 62^2 + 1984^2 = 1985^2$	”
36	252	10836	$15^2 + 36^2 + 252^2 + 10836^2 = 10839^2$	$5^2 + 12^2 + 84^2 + 3612^2 = 3613^2$	”

etc. //

Theorem 4.6. For $a = 18$ and Eq. (4.2), there hold the identities:

b	c	x	Identity	Equivalently	Ref.
3	54	540	$18^2 + 3^2 + 54^2 + 540^2 = 543^2$	$6^2 + 1^2 + 18^2 + 180^2 = 181^2$	Th. 2.1

9	66	792	$18^2+9^2+66^2+792^2 = 795^2$	$6^2+3^2+22^2+264^2=265^2$	Th. 2.3
15	90	1440	$18^2+15^2+90^2+1440^2 = 1443^2$	$6^2+5^2+30^2+480^2=481^2$	Th. 2.5
21	126	2772	$18^2+21^2+126^2+2772^2=2775^2$	$6^2+7^2+42^2+924^2=925^2$	Th. 2.6
27	174	5220	$18^2+27^2+174^2+5220^2=5223^2$	$6^2+9^2+58^2+1740^2=1741^2$	”
33	234	9360	$18^2+33^2+234^2+9360^2=9363^2$	$6^2+11^2+78^2+3120^2=3121^2$	”

etc. //

Theorem 4.7. For $a = 21$ and Eq. (4.2), there hold the identities:

b	c	x	Identity	Equivalently	Ref.
6	78	1092	$21^2+6^2+78^2+1092^2=1095^2$	$7^2+2^2+26^2+364^2 = 365^2$	Th. 2.2
12	96	1632	$21^2+12^2+96^2+1632^2=1635^2$	$7^2+4^2+32^2+544^2=545^2$	Th. 2.4
18	126	2772	$21^2+18^2+126^2+2772^2=2775^2$	$7^2+6^2+42^2+924^2=925^2$	Th. 2.6
24	168	4872	$21^2+24^2+168^2+4872^2=4875^2$	$7^2+8^2+56^2+1624^2=1625^2$	Th. 2.7
30	222	8436	$21^2+30^2+222^2+8436^2=8439^2$	$7^2+10^2+74^2+2812^2=2813^2$	”
36	288	14112	$21^2+36^2+288^2+14112^2=14115^2$	$7^2+12^2+96^2+4704^2=4705^2$	”

etc. //

Theorem 4.8. For $a = 24$ and Eq. (4.2), there hold the identities:

b	c	x	Identity	Equivalently	Ref.
3	96	1632	$24^2+3^2+96^2+1632^2=1635^2$	$8^2+1^2+32^2+544^2=545^2$	Th. 2.1

9	108	2052	$24^2 + 9^2 + 108^2 + 2052^2 = 2055^2$	$8^2 + 3^2 + 36^2 + 684^2 = 685^2$	Th. 2.3
15	132	3036	$24^2 + 15^2 + 132^2 + 3036^2 = 3039^2$	$8^2 + 5^2 + 44^2 + 1012^2 = 1013^2$	Th. 2.5
21	168	4872	$24^2 + 21^2 + 168^2 + 4872^2 = 4875^2$	$8^2 + 7^2 + 56^2 + 1624^2 = 1625^2$	Th. 2.7
27	216	7992	$24^2 + 27^2 + 216^2 + 7992^2 = 7995^2$	$8^2 + 9^2 + 72^2 + 2664^2 = 2665^2$	Th. 2.8
33	276	12972	$24^2 + 33^2 + 276^2 + 12972^2 = 12975^2$	$8^2 + 11^2 + 92^2 + 4324^2 = 4325^2$	"

etc. //

Theorem 4.9. For $a = 27$ and Eq. (4.2), there hold the identities:

b	c	x	Identity	Equivalently	Ref.
6	126	2772	$27^2 + 6^2 + 126^2 + 2772^2 = 2775^2$	$9^2 + 2^2 + 42^2 + 924^2 = 925^2$	Th. 2.2
12	144	3600	$27^2 + 12^2 + 144^2 + 3600^2 = 3603^2$	$9^2 + 4^2 + 48^2 + 1200^2 = 1201^2$	Th. 2.4
18	174	5220	$27^2 + 18^2 + 174^2 + 5220^2 = 5223^2$	$9^2 + 6^2 + 58^2 + 1740^2 = 1741^2$	Th. 2.6
24	216	7992	$27^2 + 24^2 + 216^2 + 7992^2 = 7995^2$	$9^2 + 8^2 + 72^2 + 2664^2 = 2665^2$	Th. 2.8
30	270	12420	$27^2 + 30^2 + 270^2 + 12420^2 = 12423^2$	$9^2 + 10^2 + 90^2 + 4140^2 = 4141^2$	Th. 2.9
36	336	19152	$27^2 + 36^2 + 336^2 + 19152^2 = 19155^2$	$9^2 + 12^2 + 112^2 + 6384^2 = 6385^2$	"

etc. //

Theorem 4.10. For $a = 30$ and Eq. (4.2), there hold the identities:

b	c	x	Identity	Equivalently	Ref.
3	150	3900	$30^2 + 3^2 + 150^2 + 3900^2 = 3903^2$	$10^2 + 1^2 + 50^2 + 1300^2 = 1301^2$	Th. 2.1

9	162	4536	$30^2 + 9^2 + 162^2 + 4536^2 = 4539^2$	$10^2 + 3^2 + 54^2 + 1512^2 = 1513^2$	Th. 2.3
15	186	5952	$30^2 + 15^2 + 186^2 + 5952^2 = 5955^2$	$10^2 + 5^2 + 62^2 + 1984^2 = 1985^2$	Th. 2.5
21	222	8436	$30^2 + 21^2 + 222^2 + 8436^2 = 8439^2$	$10^2 + 7^2 + 74^2 + 2812^2 = 2813^2$	Th. 2.7
27	270	12420	$30^2 + 27^2 + 270^2 + 12420^2 = 12423^2$	$10^2 + 9^2 + 90^2 + 4140^2 = 4141^2$	Th. 2.9
33	330	18480	$30^2 + 33^2 + 330^2 + 18480^2 = 18483^2$	$10^2 + 11^2 + 110^2 + 6160^2 = 6161^2$	Th. 2.10

etc. //

BIBLIOGRAPHY

[1] Alvarez, Sergio A.: Note on an *n*-dimensional Pythagorean theorem. http://www.cs.bc.edu/ alvarez/NDPyt.pdf

[2] Czyzewska, K.: Generalization of the Pythagorean theorem, *Demonstratio Math.* 24 (1991), nos. 1 - 2, 305 - 310.

[3] Misra, R.B. – Ameen, J.R.: Generalizations of Pythagoras theorem to polygons, *J. of Multidisciplinary Engg., Sci. & Tech.* (*JMEST*), *Berlin* 4 (2017), no. 8, 7778 - 7805.

[4] Oliverio, Paul: Self-generating Pythagorean quadruples and *n*-tuples. *Fibonacci Quarterly* 34 (1996), no. 2, 98 - 101.

www.ingramcontent.com/pod-product-compliance
Lightning Source LLC
Chambersburg PA
CBHW071705210326
41597CB00017B/2343
* 9 7 8 1 9 2 5 8 2 3 8 1 3 *